Systems & Control: Foundations & Applications

Marc A. Peters
Pablo A. Iglesias

Minimum Entropy Control for Time-Varying Systems

1997
Birkhäuser
Boston • Basel • Berlin

Marc A. Peters
Dept. of Electrical
& Computer Engineering
The Johns Hopkins University
Baltimore, MD 21218

Pablo A. Iglesias
Dept. of Electrical
& Computer Engineering
The Johns Hopkins University
Baltimore, MD 21218

Printed on acid-free paper

Birkhäuser ®

ISBN 0-8176-3972-1
ISBN 3-7643-3972-1
Typeset by the Authors in TEX.
Printed and bound by Edwards Brothers, Ann Arbor, MI.
Printed in the U.S.A.

9 8 7 6 5 4 3 2 1

Contents

Preface ix

1 Introduction **1**

1.1 Optimal control problems 2

1.2 Minimum entropy control 6

1.3 The maximum entropy principle 8

1.4 Extensions to time-varying systems 12

1.5 Organization of the book 13

2 Preliminaries **15**

2.1 Discrete-time time-varying systems 16

2.2 State-space realizations 21

2.3 Time-reverse systems 24

3 Induced Operator Norms **26**

3.1 Characterizations of the induced norm 27

3.2 Time-varying hybrid systems 30

3.2.1 Sampled continuous-time systems 31

3.2.2 Continuous-time systems with piecewise constant inputs 37

3.2.3 Hybrid feedback systems 41

3.3 Computational issues 44

4 Discrete-Time Entropy **46**

4.1 Entropy of a discrete-time time-varying system 47

4.2 Properties 49

4.2.1 Equivalence with the entropy integral 52

4.2.2 Entropy in terms of a state-space realization 53

4.3 Entropy and information theory 54
4.4 Entropy of an anti-causal system 57
4.5 Entropy and the $\mathcal{W}$-transform 61
4.6 Entropy of a non-linear system 64

5 Connections With Related Optimal Control Problems 69
5.1 Relationship with $\mathcal{H}_\infty$ control 70
5.2 Relationship with $\mathcal{H}_2$ control 73
5.3 Average cost functions 79
5.3.1 Average $\mathcal{H}_2$ cost 79
5.3.2 Average entropy 82
5.4 Time-varying risk-sensitive control 85
5.5 Problems defined on a finite horizon 91

6 Minimum Entropy Control 94
6.1 Problem statement 95
6.2 Basic results 96
6.3 Full information 100
6.3.1 Characterizing all closed-loop systems 100
6.3.2 FI minimum entropy controller 104
6.4 Full control 105
6.5 Disturbance feedforward 110
6.5.1 Characterizing all closed-loop systems 111
6.5.2 DF minimum entropy controller 114
6.6 Output estimation 116
6.7 Output feedback 118
6.8 Stability concepts 123

7 Continuous-Time Entropy 128
7.1 Classes of systems considered 129
7.2 Entropy of a continuous-time time-varying system 130
7.3 Properties 131
7.3.1 Equivalence with the entropy integral 136
7.3.2 Entropy in terms of a state-space realization 137
7.3.3 Relationship with discrete-time entropy 139
7.4 Connections with related optimal control problems 143
7.4.1 Relationship with $\mathcal{H}_\infty$ control 143
7.4.2 Relationship with $\mathcal{H}_2$ control 146
7.4.3 Relationship with risk-sensitive control 149

7.5 Minimum entropy control 150

A Proof of Theorem 6.5 **154**

B Proof of Theorem 7.21 **167**

Bibliography **172**

Notation **180**

Index **183**

Preface

Minimum entropy control has been studied extensively for linear time-invariant systems, both in the continuous-time and discrete-time cases. Controllers that satisfy a closed-loop minimum entropy condition are known to have considerable advantages over other optimal controllers. While guaranteeing an $\mathcal{H}_\infty$ norm bound, the entropy is an upper bound for the $\mathcal{H}_2$ norm of the system, and thus minimum entropy controllers provide a degree of performance sacrificed by other $\mathcal{H}_\infty$ controllers.

These advantages make it desirable to extend the theory of minimum entropy control to other settings. This is not straightforward, since the notion of entropy is defined in terms of the system's transfer function, which other systems may not admit.

In this book we provide a time-domain theory of the entropy criterion. For linear time-invariant systems, this time-domain notion of entropy is equivalent to the usual frequency domain criterion. Moreover, this time-domain notion of entropy enables us to define a suitable entropy for other classes of systems, including the class of linear time-varying systems. Furthermore, by working with this time-domain definition of the entropy we are able to gain new interpretations of the advantages of minimum entropy control. In particular we consider the connections between the time-varying minimum entropy control problem and the time-varying analogues to the $\mathcal{H}_2$, $\mathcal{H}_\infty$ and risk-sensitive control problem.

The majority of the work presented here arose as part of the first author's doctoral dissertation in the Department of Electrical and Computer Engineering at The Johns Hopkins University. We are grateful to our colleagues in the department, particularly to Jack Rugh, who has always been there to answer our questions.

The second author would also like to thank Professor Harry Dym. It was during a visit to Rehovot in May 1992 that the ideas behind this research were first conceived. We are also indebted to Abbie Feintuch who suggested a means for comparing the entropy operator to the quadratic control problem.

We are most grateful for the financial support of the National Science Foundation, under grant number ECS-9309387.

We would also like to thank the staff at Birkhäuser for their excellent cooperation.

Finally, we would like to thank our friends and families for their support and interest over the years. Above all, we are grateful to Elsbeth and Elizabeth, to whom we dedicate this book, for their patience and encouragement.

Marc A. Peters
Pablo A. Iglesias

Baltimore, Maryland
August 1996

1 Introduction

Minimum entropy control provides a means of trading off some of the features of other control problems, namely $\mathcal{H}_2$ optimal control and $\mathcal{H}_\infty$ control. For continuous-time, linear time-invariant (LTI) systems, a complete solution to the minimum entropy control problem is found in the monograph by Mustafa and Glover [64].

This book is devoted to the extension of this theory to time-varying systems. Using concepts from operator theory, we provide a time-domain interpretation for the entropy. This notion, equivalent to the frequency domain version of [64], allows us to consider more general classes of systems.

In this chapter we provide an introduction to the minimum entropy control problem. We do so by first considering two related optimal control problems: the $\mathcal{H}_2$ and $\mathcal{H}_\infty$ control problems. These two optimal control theories were proposed as means of achieving conflicting goals in the system: performance and robustness, respectively. We will show how minimum entropy control allows the designer to trade off these goals.

Finally, since the word entropy is usually associated with the fields of thermodynamics and information theory, we also present a brief introduction to the connection that exists between these notions of entropy, and the system entropy that is considered in this book.

1.1 Optimal control problems

Optimal control theory has been used since the 1950s. A goal of optimal control theory is to provide a theoretical basis from which to choose amongst a set of controllers. For example, a designer may wish to find a stabilizing controller K that minimizes an appropriate norm of the closed-loop transfer function. Two popular norms used in this procedure are the $\mathcal{H}_2$ and $\mathcal{H}_\infty$ norms.

In $\mathcal{H}_2$ control, the quadratic norm of a discrete-time transfer function G

$$\|G\|_2^2 := \frac{1}{2\pi}\int_{-\pi}^{\pi} \operatorname{tr}\big[G(e^{-i\omega})^T G(e^{i\omega})\big]\, d\omega$$

is minimized. This optimization problem has been suggested as a means of achieving optimal *performance* in the system.

If we assume that the input w is a zero mean white noise process with unit covariance, i.e. $\mathcal{E}(w_k) = 0$ and $\mathcal{E}(w_j w_k^T) = \delta_{j-k} I$, where $\mathcal{E}$ denotes mathematical expectation and δ_j is the Kronecker delta, the (squared) $\mathcal{H}_2$ norm arises as the cost function

$$\|G\|_2^2 = \lim_{N\to\infty} \frac{1}{N}\mathcal{E} \sum_{k=0}^{N-1} z_k^T z_k \tag{1.1}$$

In this stochastic context, the $\mathcal{H}_2$ optimization problem is better known as the linear quadratic Gaussian (LQG), or Wiener-Hopf-Kalman optimal control problem.

The theory of $\mathcal{H}_2$ optimal control was particularly popular during the 1960s. In part, this is because it is well suited to the control problems of the time — in particular, to those that arose in the aerospace industry. Moreover, since the theory can be extended easily to multivariable problems, it provides a means of designing controllers for more complicated systems than is possible using classical lead/lag and proportional/integral/derivative control.

Despite the early successes of the $\mathcal{H}_2$ theory, the application to practical problems was limited. In most industrial control problems, the systems are badly modeled, invalidating most of the assumptions made in the theory. Moreover, it has been demonstrated that optimal $\mathcal{H}_2$ controllers do not possess the requisite robustness that would allow them to be used when some of these assumptions are violated [30].

Zames, in his seminal work [85], suggested replacing the quadratic norm

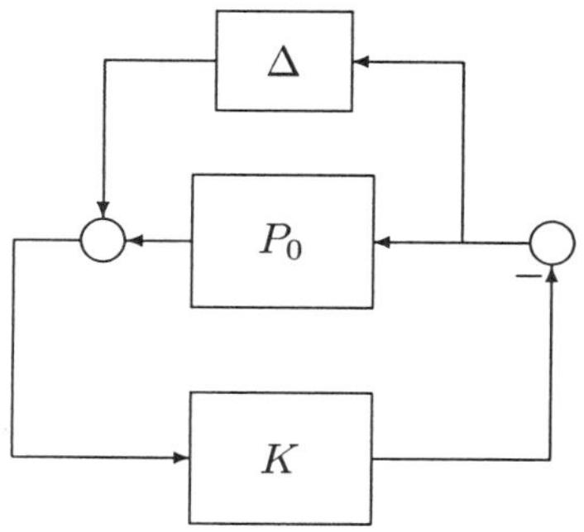

Figure 1.1: Robust control example

with the induced operator norm

$$\|G\|_\infty := \sup_{\omega\in(-\pi,\pi]} \lambda_{\max}^{1/2}\left(G(e^{-i\omega})^T G(e^{i\omega})\right)$$

known as the $\mathcal{H}_\infty$ norm of the system. Unlike the quadratic norm, the $\mathcal{H}_\infty$ norm possesses the multiplicative property ($\|GH\|_\infty \leq \|G\|_\infty \|H\|_\infty$), which allows one to guarantee some degree of robustness by use of the small gain theorem [84].

Example 1.1 Consider the system described in Figure 1.1. The nominal plant P_0 describes the model for the system that is available to the designer. The transfer function Δ models the uncertainty in the plant, and thus $P_\delta := P_0 + \Delta$ represents the actual plant. There are no assumptions made on the unknown transfer function Δ other than: 1) Δ is stable — this implies that P_0 and P_δ have the same number of unstable poles; and, 2) there is a function of frequency, say $W(e^{i\omega})$ such that

$$\left\|\Delta(e^{i\omega})\right\| < W(e^{i\omega}), \qquad \text{for } \omega \in (-\pi, \pi]$$

A controller is sought that will stabilize all the perturbed plants P_δ satisfying these two assumptions. By redrawing the system as in Figure 1.2, where

$$G := (I + KP_0)^{-1}K$$

we are able to use the small gain theorem to conclude that the system is is stable if both G and Δ are stable and $\|G\|_\infty \|\Delta\|_\infty < 1$. This will be satisfied if

$$\|WG\|_\infty =: \gamma_0 < 1$$

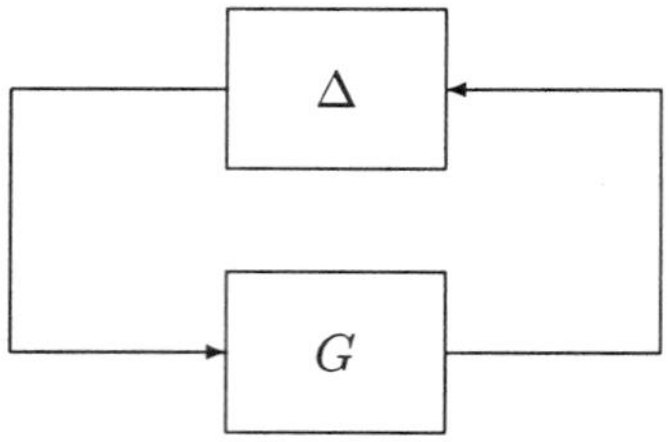

Figure 1.2: Equivalent setup for use of small gain theorem

Based on these assumptions, the maximally robust controller will be the controller K that makes G stable and provides the smallest $\mathcal{H}_\infty$ norm γ_0 for the closed-loop system G. □

The $\mathcal{H}_\infty$ norm also has a time-domain interpretation. Define the cost function

$$J := \inf_{w \in \ell_{2+}} \sum_{k=0}^{\infty} \left(w_k^T w_k - \gamma^{-2} z_k^T z_k \right) \tag{1.2}$$

Clearly

$$J > 0 \iff \gamma^2 > \sup_{0 \neq w \in \ell_{2+}} \frac{\sum_{k=0}^{\infty} z_k^T z_k}{\sum_{k=0}^{\infty} w_k^T w_k} =: \|G\|_\infty^2$$

The optimization problem implied by (1.2) is well known in the context of linear differential quadratic games [13].

Optimal control problems such as those of Example 1.1 can be recast in terms of a standard set up, called the *standard problem.* Consider the closed-loop system depicted by Figure 1.3. The generalized plant P, which is assumed to be an LTI system, includes any weighting functions (such as the function W in Example 1.1) along with the model of the plant. The disturbance input w includes any external inputs as well as any unknown disturbances. The signal u is the control input, y is the measured output and z is the output signal. The controller K is constrained such that K provides internal stability; that is, the states of P and K go to zero from any initial value when $w \equiv 0$. The closed-loop transfer function from w to

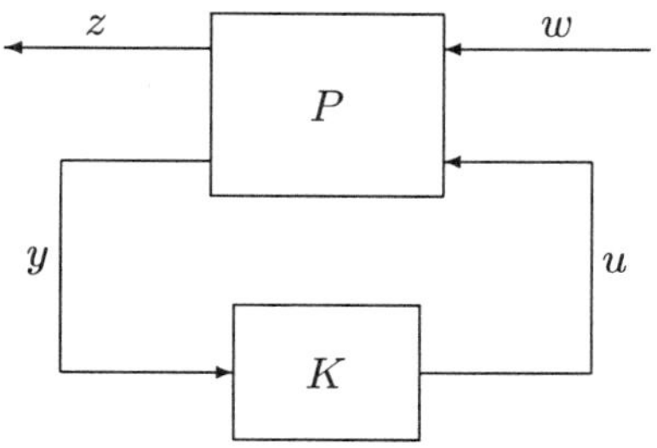

$$P = \begin{bmatrix} P_{11} & P_{12} \\ P_{21} & P_{22} \end{bmatrix}$$

Figure 1.3: Closed-loop system $G = \mathcal{F}_\ell(P, K)$

z is given by a linear fractional transformation of K and P

$$G := \mathcal{F}_\ell(P, K) := P_{11} + P_{12}K(I - P_{22}K)^{-1}P_{21} \tag{1.3}$$

Consequently, the standard $\mathcal{H}_\infty$ control problem is to find an internally stabilizing controller K such that $\|\mathcal{F}_\ell(P, K)\|_\infty$ is minimized.

Example 1.2 We can restate the feedback system of Example 1.1 in terms of the standard problem by defining

$$P := \begin{bmatrix} P_{11} & P_{12} \\ P_{21} & P_{22} \end{bmatrix} := \begin{bmatrix} 0 & W \\ I & -P_0 \end{bmatrix}$$

from which we get $\mathcal{F}_\ell(P, K) = WK(I + P_0K)^{-1}$. □

The early method of solution for the $\mathcal{H}_\infty$ control problem is outlined in the monograph by Francis [37]. The linear fractional transformation (1.3) can be simplified by the use of the Youla parameterization of all stabilizing controllers. It can be shown that all closed-loop systems can be written as $T_1 - T_2QT_3$ for some free stable parameter Q. It follows that the standard $\mathcal{H}_\infty$ problem is equivalent to the so-called *model matching problem*

$$\inf_{Q \in \mathcal{H}_\infty} \|T_1 - T_2QT_3\|_\infty$$

This problem can be further simplified to obtain the equivalent optimization problem

$$\inf_{Q \in \mathcal{H}_\infty} \left\| \begin{bmatrix} R_{11} & R_{12} \\ R_{21} + Q & R_{22} \end{bmatrix} \right\|_\infty \tag{1.4}$$

It can be shown that the resulting R_{ij} are unstable; that is $R^{\sim} \in \mathcal{H}_\infty$ [37]. A solution to this problem, known in the control community as the "four block problem," has been available for some time in the mathematics literature [26, 66]. As such, the $\mathcal{H}_\infty$ problem could be said to be solved. Nevertheless, it was not in the most convenient form for use in control theory. In particular, for a system admitting a state-space realization, the steps involved in obtaining and solving the four block problem involved a degree inflation that resulted in controllers of extremely high degree.

By considering a series of simpler problems, and through a special separation principle, Glover and Doyle and co-workers presented a complete solution to the ***sub-optimal*** $\mathcal{H}_\infty$ control problem for continuous-time LTI systems [31, 40]. That is, given a number $\gamma > 0$, necessary and sufficient conditions are given on the existence of a stabilizing controller K satisfying $\|\mathcal{F}_\ell(P, K)\|_\infty < \gamma$. Moreover, explicit state-space formulae for *all* the controllers achieving this norm bound are given involving the solution of only two algebraic Riccati equations. The resultant controller is of the same degree as the generalized plant.

The Glover-Doyle result was revolutionary on two accounts. First of all, it bypasses the use of the Youla parameterization and the four block problem. More importantly, it highlights the connection between $\mathcal{H}_\infty$ control and $\mathcal{H}_2$ control. This time-domain derivation of the $\mathcal{H}_\infty$ control also makes it amenable for extensions to settings beyond LTI systems, including time-varying, infinite-dimensional and non-linear systems.

1.2 Minimum entropy control

Sub-optimal controllers that guarantee that the $\mathcal{H}_\infty$ norm of the closed-loop operator is below a prescribed bound are not unique. In fact, all controllers achieving this norm bound can be expressed in terms of a linear fractional transformation of a fixed controller K_c and a free parameter with an induced norm constraint. That is, the set of all controllers can be expressed as

$$K = \mathcal{F}_\ell(K_c, Q); \qquad Q \in \mathcal{H}_\infty, \quad \|Q\|_\infty < 1$$

Since the (1,1) block of the controller $(K_c)_{11}$ is in some sense at the center $(Q = 0)$ of the ball of all controllers, it is known as the *central* controller.

While every choice of this parameter will guarantee the same closed-loop norm bound, it is useful to ask whether there exists a choice that will minimize some auxiliary, desirable closed-loop performance measure.

A natural choice for this auxiliary cost is the quadratic norm of the closed-loop transfer function. One of the drawbacks of optimal $\mathcal{H}_\infty$ control is that the performance that is usually associated with $\mathcal{H}_2$ control is sacrificed in favor of $\mathcal{H}_\infty$ control's superior robustness. In an engineering context, neither of these two extremes is desired. A means of combining the two objectives as described above is preferable.

An approach for trading off robustness and performance in a design was suggested by Mustafa and Glover, who proposed a sub-optimal $\mathcal{H}_\infty$ control problem which minimizes the *entropy* of the closed-loop system. For a discrete-time transfer function G with $\|G\|_\infty < \gamma$, the entropy is defined by

$$I(G,\gamma,z_0) := -\frac{\gamma^2}{2\pi}\int_{-\pi}^{\pi} \ln\left|\det\left(I-\gamma^{-2}G(e^{-i\omega})^T G(e^{i\omega})\right)\right| \frac{|z_0|^2-1}{|z_0-e^{i\omega}|^2}d\omega$$

for some $z_0 \in \mathbb{C}$, $|z_0| > 1$. Typically, we take $z_0 \uparrow \infty$, where the limit exists by the dominated convergence theorem, resulting in

$$I(G,\gamma) := -\frac{\gamma^2}{2\pi}\int_{-\pi}^{\pi} \ln\left|\det\left(I-\gamma^{-2}G(e^{-i\omega})^T G(e^{i\omega})\right)\right| d\omega \qquad (1.5)$$

In their research on minimum entropy control for continuous-time, LTI systems, Mustafa and Glover have shown that controllers based on this paradigm have considerable advantages over other sub-optimal $\mathcal{H}_\infty$ controllers. Similar results for discrete-time systems have been obtained by Iglesias and Mustafa [53, 54].

For LTI systems, it can been shown that the entropy is an upper bound on the squared quadratic norm of the system. A power series expansion yields

$$-\gamma^2\ln\left|\det\left(I-\gamma^{-2}G(e^{-i\omega})^T G(e^{i\omega})\right)\right| \geq \operatorname{tr}\left[G(e^{-i\omega})^T G(e^{i\omega})\right]$$

from which $I(G,\gamma) \geq \|G\|_2^2$ follows. Hence, minimizing the entropy will ensure that the closed-loop quadratic norm will not be overly large. Note that it is possible to choose $\mathcal{H}_\infty$ sub-optimal controllers which do not exhibit this property. In fact, in continuous-time it is possible to select $\mathcal{H}_\infty$ sub-optimal controllers which have an unbounded closed-loop quadratic norm. Furthermore, if the bound on the $\mathcal{H}_\infty$ norm is relaxed completely, the entropy of a system turns out to be equal to the squared quadratic norm. This follows by rewriting

$$-\gamma^2\ln\left|\det\left(I-\gamma^{-2}G(e^{-i\omega})^T G(e^{i\omega})\right)\right| = \operatorname{tr}\left[G(e^{-i\omega})^T G(e^{i\omega})\right] + \mathcal{O}(\gamma^{-2})$$

and letting γ tend to infinity.

Minimum entropy controllers have also been shown to be equivalent to the *risk-sensitive* control problem — also known as the linear exponential quadratic Gaussian control problem (LEQG). The risk-sensitive optimal controller minimizes

$$R_N(\theta) = -\frac{2}{\theta N} \ln \mathcal{E} \exp\left(-\frac{\theta}{2} \sum_{k=0}^{N-1} z_k^T z_k\right)$$

where the input is Gaussian white noise. In [40], Glover and Doyle showed that, for $\|G\|_\infty < \gamma$, the two problems are connected by the following relationship:

$$\lim_{N \to \infty} R_N(-\gamma^{-2}) = I(G, \gamma)$$

The definition of the entropy used to solve this optimal control control problem is related to other common notions of entropy found in thermodynamics and information theory. In the next section we analyze this connection.

1.3 The maximum entropy principle

The entropy postulate of thermodynamics states that the entropy will increase to the maximum allowed by the constraints governing the system. In statistical thermodynamics, it was Gibbs who showed that this postulate amounts to defining the theoretical entropy S as

$$\mathcal{S} = -k_B \sum p_i \ln p_i \tag{1.6}$$

where p_i is the probability that a particle is at a given energy state, and k_B is Boltzmann's constant. The sum is carried out over all thermodynamic microstates that are consistent with the imposed macroscopic constraints. It can be shown that the microstates of a system will be distributed according to the distribution $p_i^{\max}$ which maximizes the theoretical entropy. With this distribution, the theoretical entropy (1.6) will be numerically equal to the experimental entropy defined by Clausius [44].

The use of (1.6) as a measure of "information" in a random signal was proposed by Hartley [47]. This notion of entropy now forms the corner stone of information theory, as formulated by Shannon [73]. Shannon proposed a means of defining a single numerical measure which would give the amount of uncertainty represented by a probability distribution. Specifically, for a discrete random variable with probability density function p_i, the entropy, or amount of *information*, is defined as

$$\mathcal{H} = -\sum p_i \ln p_i \tag{1.7}$$

Clearly, up to a constant k_B, the entropies defined by (1.6) and (1.7) are equivalent. It was Jaynes who established the relationship between these two notions of entropy [58]. More importantly, he used this connection to propose a means of assigning unknown probabilities to problems where there is incomplete or noisy data. This principle is illustrated in the choice made by nature to assign the probability of the microstates so as to maximize the entropy subject to the imposed constraints implied by the data. Today, this maximum entropy principle is used in a variety of fields.

In the following example we provide a very simple application of the maximum entropy principle.

Example 1.3 Consider a zero-mean, stationary stochastic process x_i whose n-step autocorrelation matrix R is given by

$$R := \mathcal{E}\left\{\boldsymbol{x}\boldsymbol{x}^T\right\} > 0$$

where

$$\boldsymbol{x} := \begin{bmatrix} x_0 \\ x_1 \\ \vdots \\ x_{n-1} \end{bmatrix}$$

Suppose that you wish to assign a probability density function $f(\boldsymbol{x})$ to the process, consistent with the available data. Since this choice is not unique, we can use the maximum entropy principle as a basis for making this choice. In particular, it is easy to show that the density which maximizes the entropy (1.7) is the Gaussian distribution

$$f(\boldsymbol{x}) = \frac{1}{\sqrt{(2\pi)^n \det R}} \exp\left(-\frac{1}{2}\boldsymbol{x}^T R^{-1} \boldsymbol{x}\right)$$

Moreover, the entropy equals

$$\mathcal{H}(\boldsymbol{x}) = \frac{n}{2} \ln 2\pi e + \frac{1}{2} \ln \det R \tag{1.8}$$

Note that in this case, we have used the analogue of (1.7) for continuous-type random variables. □

Remark 1.4 The matrix R in the example above is a Toeplitz matrix. Using limits derived by Szegö [43] on the distribution of the eigenvalues of such matrices, it can be shown that the limit of the *entropy rate* of this process is

$$\bar{\mathcal{H}} := \lim_{n\to\infty} \frac{1}{n}\mathcal{H}(\boldsymbol{x}) = \frac{1}{2}\ln 2\pi e + \frac{1}{2\sigma}\int_{-\sigma}^{\sigma} \ln S(e^{i\omega})d\omega$$

where $S(z)$ is the power spectrum of the bandlimited Gaussian process $\boldsymbol{x}$.

The problem described in Example 1.3 is related to the minimum entropy control problem described above. To show this connection, we first consider a matrix interpolation problem.

Example 1.5 Define the following $n \times n$ real matrix.

$$H := \begin{bmatrix} h_0 & \cdots & h_k & 0 & \cdots & \cdots & 0 \\ \vdots & & & \ddots & 0 & & \vdots \\ h_k & & & & \ddots & \ddots & \vdots \\ 0 & h_k & & & & \ddots & 0 \\ \vdots & 0 & \ddots & & & & h_k \\ \vdots & & \ddots & \ddots & & & \vdots \\ 0 & \cdots & \cdots & 0 & h_k & \cdots & h_0 \end{bmatrix}$$

Suppose that we wish to find an extension to the matrix H; i.e. a matrix Q with $q_{i,j} = 0$ for $|i-j| < k$,y such that the matrix $R := H + Q$ is positive definite. Denote by $\mathcal{Q}(H)$ the set of all Q achieving this constraint. If this set is non empty, then we wish to choose the matrix Q_o for which

$$\det R_o \leq \det R, \qquad \text{for all } Q \in \mathcal{Q}(H)$$

where $R_o := H + Q_o$. It can be shown, for example [33], that there indeed exists such a Q; moreover, it is the unique $Q \in \mathcal{Q}(H)$ such that R^{-1} exhibits the same band structure that H does. □

The matrix extension problem considered above can now be used to generalize the problem considered in Example 1.3.

Example 1.6 Consider again Example 1.3, but assume that in this case, not all of the correlation matrix is available. In fact, we assume that only the first $k+1$ diagonals are known. In other words, the correlation matrix has the band structure of H in Example 1.5. We treat the zero elements as unknowns to be filled. For this matrix to be the correlation matrix of a normal distribution, it must be positive definite. Moreover, we choose the distribution according to the maximum entropy principle. From the value of the entropy given by (1.8), and using the fact that the logarithm is a monotonic function, it follows that to maximize the entropy, the optimal choice of Q is the Q_o mentioned in Example 1.5. □

Remark 1.7 The matrix extension problem outlined in 1.5 was first considered by Chover [22]. Nevertheless, it is usually associated with Burg, who first used it to solve the discrete-time covariance extension problem outlined in Example 1.6 [18].

The problems considered in Examples 1.3, 1.5 and 1.6 are classical uses of the maximum entropy principle. Moreover, they can be used to pose the minimum entropy control problem as an example of the maximum entropy principle.

Recall that the standard problem can be restated in terms of the four block problem, where we wish to find a $Q \in \mathcal{H}_\infty$ according to (1.4). Since the $R_{ij}(z)$ are anti-causal, we can associate with each of these blocks an infinite-dimensional upper triangular Toeplitz matrix $\boldsymbol{R}$, where the ith diagonal is just the ith term of the impulse response of R. Similarly, for the stable extension $Q(z)$, we associate a lower triangular Toeplitz matrix $\boldsymbol{Q}$.

In terms of these matrices, the four block problem can be rewritten as

$$(1.4) < \gamma \quad \Longleftrightarrow \quad \gamma^2 \boldsymbol{I} - \begin{bmatrix} \boldsymbol{R}_{11} & \boldsymbol{R}_{12} \\ \boldsymbol{R}_{21} + \boldsymbol{Q} & \boldsymbol{R}_{22} \end{bmatrix}^* \begin{bmatrix} \boldsymbol{R}_{11} & \boldsymbol{R}_{12} \\ \boldsymbol{R}_{21} + \boldsymbol{Q} & \boldsymbol{R}_{22} \end{bmatrix} > \boldsymbol{0}$$

$$\Longleftrightarrow \quad \begin{bmatrix} \gamma \boldsymbol{I} & \boldsymbol{0} & \boldsymbol{R}_{11}^* & \boldsymbol{R}_{21}^* + \boldsymbol{Q}^* \\ \boldsymbol{0} & \gamma \boldsymbol{I} & \boldsymbol{R}_{12}^* & \boldsymbol{R}_{22}^* \\ \boldsymbol{R}_{11} & \boldsymbol{R}_{12} & \gamma \boldsymbol{I} & \boldsymbol{0} \\ \boldsymbol{R}_{21} + \boldsymbol{Q} & \boldsymbol{R}_{22} & \boldsymbol{0} & \gamma \boldsymbol{I} \end{bmatrix} > 0 \qquad (1.9)$$

It follows that the matrix in (1.9) exhibits the same band structure as the matrix H in the band extension problem. Moreover, so does the extension $\boldsymbol{Q}$. As such, the minimum entropy control problem is equivalent to the infinite-dimensional analogue of the band extension problem considered in Example 1.5.

The association of a causal system with a lower triangular matrix is central to the development of the entropy that will be used in this monograph. This association will be made explicit in Section 2.1. In the next section we discuss the extension of the minimum entropy control theory to time-varying systems.

1.4 Extensions to time-varying systems

We have shown that there is a significant practical motivation for choosing controllers which have entropy as an auxiliary cost. Generalizations of the Riccati equation based solution to the $\mathcal{H}_\infty$ sub-optimal control problem to more general classes of systems have now appeared in the literature. Nevertheless, generalizations of the minimum entropy control paradigm have been more elusive, since the definition of the entropy of a system is intimately tied to the system's transfer function. For this reason an extension to classes of systems beyond LTI systems is not straightforward.

A definition for the entropy of a discrete-time, linear time-varying (LTV) systems was proposed by Iglesias [51]. This generalization exploited the connection between the entropy of the closed-loop transfer function and the spectral factorization problem. The entropy was defined in terms of the $\mathcal{W}$-transform of the system's spectral factor. The $\mathcal{W}$-transform, introduced by Alpay *et al.* [1] in the context of interpolation problems for non-stationary processes, is a generalization of the usual $\mathcal{Z}$-transform to non-Toeplitz operators.

In this book we build upon the definition introduced in [51] to provide a complete theory of minimum entropy control for linear time-varying systems. We first provide a slightly modified definition of the entropy, which is consistent with that used for time-invariant systems. We then investigate properties of this entropy, in particular its relationship with the optimal control problems mentioned in this chapter. Finally, we solve the minimum entropy control problem for systems admitting a state-space realization. Note that the entropy defined in [51] for control systems is related to the entropy used by Gohberg *et al.* [41] who solved a generalization of the band extension problem of Example 1.5 to non-Toeplitz operators. Owing to the equivalence between the $\mathcal{H}_\infty$ control problem and the band extension problem considered in Section 1.3, the minimum entropy control problem could be solved by using the results of [41]. Nevertheless, as was done for the LTI case, we will follow the approach introduced by Arov and Kreĭn [6, 7] who first used linear fractional transformations to solve a series of minimum entropy extension problems.

1.5 Organization of the book

The rest of this book is organized as follows:

Chapter 2: Preliminaries

In this chapter we introduce the class of systems that will be considered in the rest of the book. We will be primarily interested in discrete-time time-varying systems. These systems will be represented using infinite-dimensional operators, formalizing the approach used in Section 1.3. Moreover, we present preliminary material that will be central to the development that will follow.

Chapter 3: Induced Operator Norms

For the entropy (1.5) to make sense, we require that the $\mathcal{H}_\infty$ norm of the closed-loop system G be bounded by γ. For time-varying systems this translates into the requirement that the induced norm of the system be bounded. In this chapter we provide characterizations for the induced norms of two types of systems. It is first shown that the induced ℓ_2 norm of a discrete-time linear time-varying system may be characterized by the existence requirement on solutions to operator Riccati equations. A similar result is derived for systems that arise in sampled-data systems involving a mixture of continuous- and discrete-time signals.

Chapter 4: Discrete-Time Entropy

In this chapter we present our definition of the entropy for the discrete-time time-varying systems. Unlike time-invariant systems, where one definition suffices, it is necessary to define entropies for both causal and anti-causal systems. Some basic properties of the entropy are also investigated. In particular, we consider the connection with the entropy used in information theory. Moreover, the possible extension of our notion of entropy to non-linear systems is considered.

Chapter 5: Connections With Related Optimal Control Problems

We have shown above that one of the primary motivations for the use of minimum entropy controllers is the connection with several related optimal control problems. In this chapter we consider the analogous properties for linear time-varying systems. In particular, the relationship between the time-varying entropy introduced here and related $\mathcal{H}_\infty$ and $\mathcal{H}_2$ cost functions is considered. Moreover, the relationship between the entropy of a causal system and the entropy of its anti-causal adjoint is demonstrated. This relationship is used to compare the entropy operator and a related risk-sensitive cost function.

Chapter 6: Minimum Entropy Control

In this chapter we pose the problem of finding a controller which minimizes the entropy of the closed-loop input-output operator for discrete-time time-varying systems. These controllers are of considerable practical significance because they provide a simple means of trading off robustness and performance. We do so by first providing a characterization of all stabilizing controllers which satisfy an induced-norm bound. This characterization is derived along the lines used by Glover and Doyle in their solution of the $\mathcal{H}_\infty$ control problem [31]. The minimum entropy control problem is then solved via a separation principle.

Chapter 7: Continuous-Time Entropy

In this final chapter we present a definition of the entropy for time-varying continuous-time systems. This entropy differs significantly from the entropy for discrete-time systems. Properties of this entropy, including the connections with related optimal control problems, as well as the connection with the discrete-time entropy, are discussed.

Preliminaries 2

In this chapter we introduce the class of systems that will be considered, as well as the notation used throughout the book. We also introduce several basic concepts, and present several preliminary results.

Throughout the book, we will be dealing primarily with linear discrete-time time-varying systems. These systems will be represented as infinite-dimensional operators. This set-up enables us to define a time-domain based notion of entropy, which will be outlined in Chapter 4. Before doing so, we discuss some general properties of these input-output operators. The non-compactness of the operators requires us to reformulate several notions of finite-dimensional matrix theory to the infinite-dimensional setting.

Next we restrict our attention to a special class of systems, namely those which can be represented by a state-space realization. We will outline the notion of stability for such systems, and give some preliminary results regarding stability properties associated with the state-space operators governing the realization of the system.

Finally, we will consider the concept of duality for time-varying systems. The connection between a causal system and its anti-causal adjoint is not straightforward in the time-varying case. To investigate this connection we introduce the so-called time-reverse operator. This operator will allow us to compare these two classes of systems.

2.1 Discrete-time time-varying systems

In this section we provide a characterization of discrete-time, time-varying systems using infinite-dimensional operators. We first recall some basic facts about discrete-time systems and signals that are needed in the sequel.

The notation is more or less standard. All matrices and vectors are assumed real. For a matrix $M \in \mathbb{R}^{m \times n}$, M^T denotes its transpose. We consider the set of sequences from $\mathbb{Z} \to \mathbb{R}^n$. The subset consisting of square summable sequences is denoted ℓ_2^n. This is a Hilbert space with inner product

$$\langle \boldsymbol{x}, \boldsymbol{y} \rangle_{\ell_2^n} := \sum_{k=-\infty}^{\infty} x_k^T y_k$$

and norm

$$|\boldsymbol{x}|_2 := \sqrt{\sum_{k=-\infty}^{\infty} |x_k|^2} < \infty$$

Where the dimension of the underlying space is immaterial this will be omitted.

A linear operator $\boldsymbol{G}$ has a matrix representation

$$\boldsymbol{G} = \begin{bmatrix} \ddots & \vdots & \vdots & \iddots \\ \cdots & G_{0,0} & G_{0,1} & \cdots \\ \cdots & G_{1,0} & G_{1,1} & \cdots \\ \iddots & \vdots & \vdots & \ddots \end{bmatrix}$$

In the space of linear operators we designate the subset of bounded operators $\mathcal{B}(\ell_2^m, \ell_2^p)$ mapping ℓ_2^m to ℓ_2^p. We will often omit the spaces the operator is working on, since this is clear from the context. The subspace of $\mathcal{B}$ of causal operators is denoted $\mathcal{C}$, and the memoryless operators $\mathcal{M}$. If $\boldsymbol{G}$ is causal, then $G_{k,l} = 0$ for $k < l$; if $\boldsymbol{G}$ is anti-causal, then $G_{k,l} = 0$ for $k > l$. The memoryless operators are those with block-diagonal matrix representations, and for those operators we will in general use the shorter notation $\boldsymbol{G} = \text{diag}\{G_k\}$.

By the adjoint of an operator $\boldsymbol{G} : \ell_2^m \to \ell_2^p$ we denote the unique operator $\boldsymbol{G}^* : \ell_2^p \to \ell_2^m$ satisfying

$$\langle \boldsymbol{x}, \boldsymbol{G}\boldsymbol{y} \rangle_{\ell_2^p} = \langle \boldsymbol{G}^*\boldsymbol{x}, \boldsymbol{y} \rangle_{\ell_2^m}$$

for all $\boldsymbol{x} \in \ell_2^p$ and $\boldsymbol{y} \in \ell_2^m$. For the operators considered here, the matrix representing the adjoint is just the transpose of the infinite-dimensional matrix $\boldsymbol{G}$.

By the norm of an operator we mean the ℓ_2 induced operator norm

$$\|\boldsymbol{G}\| = \sup_{0 \neq \boldsymbol{w} \in \ell_2^m} \frac{\|\boldsymbol{G}\boldsymbol{w}\|_2}{\|\boldsymbol{w}\|_2}$$

For bounded operators, the norms of individual components can be bounded by norm of the operator. That is:

Lemma 2.1 *For $\boldsymbol{G} \in \mathcal{B}$, we have $\|G_{i,j}\| \leq \|\boldsymbol{G}\|$.*

Proof: Define the operator

$$\boldsymbol{E}_i := \begin{bmatrix} \vdots \\ 0 \\ I \\ 0 \\ \vdots \end{bmatrix}$$

where the identity matrix is in the ith block. It follows that

$$\|G_{i,j}\| = \|\boldsymbol{E}_i^* \boldsymbol{G} \boldsymbol{E}_j\| \leq \|\boldsymbol{E}_i\| \, \|\boldsymbol{G}\| \, \|\boldsymbol{E}_j\| \leq \|\boldsymbol{G}\|$$

as required. ■

For diagonal operators, an explicit expression for the norm is possible.

Lemma 2.2 *For $\boldsymbol{G} \in \mathcal{M}$,*

$$\|\boldsymbol{G}\| = \sup_i \|G_i\|$$

Proof: For the proof of this we follow [1]. Let $\gamma := \sup_i \|G_i\|$ which is bounded by Lemma 2.1. Then, for any $\boldsymbol{w} \in \ell_2$

$$\begin{aligned} \|\boldsymbol{G}\boldsymbol{w}\|_2^2 &= \sum_{i=-\infty}^{\infty} \|G_i w_i\|^2 \\ &\leq \gamma^2 \sum_{i=-\infty}^{\infty} |w_i|^2 \\ &= \gamma^2 \|\boldsymbol{w}\|^2 \end{aligned}$$

Hence, $\|\boldsymbol{G}\| \leq \gamma$. Now, suppose that $\boldsymbol{G} \neq \boldsymbol{0}$ — otherwise, the assertion is obvious. From Lemma (2.1), we know that for every integer i, $\|G_i\| \leq \|\boldsymbol{G}\|$. It follows that for any $\epsilon > 0$ there exists an index i such that

$$\|G_i\| \geq \gamma - \epsilon$$

For this choice of i, there exists a unit vector $\bar{w}$ such that

$$\|G_i\bar{w}\| \geq \|G_i\| - \epsilon$$

Defining the input sequence $\boldsymbol{w} = \{w_j\}$ according to:

$$w_j = \begin{cases} \bar{w} & \text{for } j = i \\ 0 & \text{elsewhere} \end{cases}$$

we have that

$$\|\boldsymbol{G}\boldsymbol{w}\| = \|G_i\bar{w}\| \geq \gamma - 2\epsilon$$

which implies that

$$\|\boldsymbol{G}\| = \|G_i\| \geq \gamma - 2\epsilon$$

and hence, since ϵ is an arbitrary positive number, the proof is complete. ■

An operator $\boldsymbol{T} \in \mathcal{C}$ is said to be *inner* if it is norm preserving, i.e. $\boldsymbol{T}^*\boldsymbol{T} = \boldsymbol{I}$, and *co-inner* if $\boldsymbol{T}\boldsymbol{T}^* = \boldsymbol{I}$.

The operator $\boldsymbol{T} \in \mathcal{B}$ mapping ℓ_2^n to ℓ_2^n is called invertible, denoted $\boldsymbol{T} \in \mathcal{B}^{-1}$, if there exists an operator $\boldsymbol{S} \in \mathcal{B}$ mapping ℓ_2^n to ℓ_2^n such that $\boldsymbol{S}\boldsymbol{T} = \boldsymbol{T}\boldsymbol{S} = \boldsymbol{I}$. The operator $\boldsymbol{S}$ is the inverse of $\boldsymbol{T}$, denoted $\boldsymbol{S} = \boldsymbol{T}^{-1}$. Similarly we say that $\boldsymbol{T} \in \mathcal{C}^{-1}$ (respectively $\boldsymbol{T} \in \mathcal{M}^{-1}$) if $\boldsymbol{T}^{-1} \in \mathcal{C}$ (respectively $\boldsymbol{T}^{-1} \in \mathcal{M}$).

We say that an operator $\boldsymbol{W} : \ell_2^n \to \ell_2^n$ is positive definite ($\boldsymbol{W} > \boldsymbol{0}$) if $\boldsymbol{W} = \boldsymbol{W}^*$ and if there exists an $\epsilon > 0$ such that $\langle \boldsymbol{x}, \boldsymbol{W}\boldsymbol{x} \rangle \geq \epsilon \langle \boldsymbol{x}, \boldsymbol{x} \rangle$ for all $\boldsymbol{x} \in \ell_2^n$. The definition of an operator being positive semi-definite is obvious. It can be checked that, if $\boldsymbol{W} > \boldsymbol{0}$ is bounded, it has a bounded positive definite inverse.

A causal operator $\boldsymbol{M}$ can be decomposed as $\boldsymbol{M} = \boldsymbol{M}_m + \boldsymbol{M}_{sc}$, where $\boldsymbol{M}_m$ is the memoryless part, and $\boldsymbol{M}_{sc}$ is the strictly causal part. For this memoryless part we have a preliminary result.

Lemma 2.3 *Suppose that $\boldsymbol{M} \in \mathcal{C} \cap \mathcal{C}^{-1}$. Then $\boldsymbol{M}_m \in \mathcal{M} \cap \mathcal{M}^{-1}$.*

Proof: First note that the memoryless part of the operator satisfies

$$\|\boldsymbol{M}_m\| = \sup_i \|M_{i,i}\| \leq \|\boldsymbol{M}\|$$

by Lemmas 2.2 and 2.1 respectively. It follows that the memoryless part is a bounded operator. Since $\boldsymbol{M}$ is invertible, there exists an $\boldsymbol{R} \in \mathcal{C}$ such that $\boldsymbol{MR} = \boldsymbol{RM} = \boldsymbol{I}$. Since both operators are causal, it follows that $\boldsymbol{M}_m\boldsymbol{R}_m = \boldsymbol{R}_m\boldsymbol{M}_m = \boldsymbol{I}$. It is easy to check, as above, that $\boldsymbol{R}_m$ is bounded, since $\boldsymbol{R}$ is causal and bounded. Hence $\boldsymbol{M}_m$ has a memoryless inverse and $\boldsymbol{M}_m^*\boldsymbol{M}_m > \boldsymbol{0}$. ∎

Throughout the sequel, we will make ample use of the *forward shift* operator $\boldsymbol{Z}$. It is defined as the operator satisfying

$$(\boldsymbol{Zx})_k := x_{k+1}$$

We will also be using two projections. The first, denoted $\boldsymbol{P}_k$, is the operator for which

$$\boldsymbol{P}_k\boldsymbol{x} := \begin{bmatrix} \vdots \\ x_{k-1} \\ x_k \\ 0 \\ \vdots \end{bmatrix}$$

The second projection, denoted $\boldsymbol{P}_k^{\perp}$, is just the orthogonal complement of the first; i.e.

$$\boldsymbol{P}_k^{\perp} := \boldsymbol{I} - \boldsymbol{P}_k$$

Notice that $\boldsymbol{Z}$ is an invertible operator on ℓ_2 with inverse $\boldsymbol{Z}^*$. In fact, when acting on ℓ_2, it is both an inner and co-inner operator. Note that if the space upon which the operator acts on is the space of singly infinite square-summable sequences, i.e. ℓ_{2+}, then $\boldsymbol{Z}$ is co-inner, but it is *not* inner. Furthermore, it is straightforward to write down a representation of these three operators.

Throughout the sequence, we will require the notion of spectral factors and co-spectral factors of positive definite operators. In system theory, the spectral factorization problem is an old one; early solutions are due to Youla [82]. The form of the result that is used here is due to Arveson [8]. It provides an existence proof for the decomposition of a positive-definite operator into the product of a causal operator and its adjoint. As such, this can be seen as a generalization of the LU decomposition familiar in finite-dimensional linear algebra.

Lemma 2.4 ([8]) *Suppose that* $\boldsymbol{W} \in \mathcal{B} \cap \mathcal{B}^{-1}$ *is a positive definite operator. There exists a spectral factorization of* $\boldsymbol{W}$: *an operator* $\boldsymbol{V} \in \mathcal{C} \cap \mathcal{C}^{-1}$ *such that*

$$\boldsymbol{W} = \boldsymbol{V}^*\boldsymbol{V}$$

Moreover, if $\boldsymbol{S}$ *is also a spectral factor of* $\boldsymbol{W}$*, then* $\boldsymbol{S} = \boldsymbol{U}\boldsymbol{V}$ *for some* $\boldsymbol{U} \in \mathcal{M} \cap \mathcal{M}^{-1}$ *with* $\boldsymbol{U}^*\boldsymbol{U} = \boldsymbol{I}$.

Similarly there exists a co-spectral factorization of $\boldsymbol{W}$*: this is an operator* $\bar{\boldsymbol{V}} \in \mathcal{C} \cap \mathcal{C}^{-1}$ *such that*

$$\boldsymbol{W} = \bar{\boldsymbol{V}}\bar{\boldsymbol{V}}^*$$

Moreover, if $\bar{\boldsymbol{S}}$ *is also a co-spectral factor of* $\boldsymbol{W}$*, then* $\bar{\boldsymbol{S}} = \bar{\boldsymbol{V}}\bar{\boldsymbol{U}}$ *for some* $\bar{\boldsymbol{U}} \in \mathcal{M} \cap \mathcal{M}^{-1}$ *with* $\bar{\boldsymbol{U}}\bar{\boldsymbol{U}}^* = \boldsymbol{I}$. □

Since our operators are non-compact, much care is needed in existence and convergence statements. However, we will often talk about block diagonal operators (i.e. memoryless operators), for which concepts follow easily as a result of standard matrix theory. Some examples: a memoryless operator $\boldsymbol{0} \le \boldsymbol{W}$ can be factored as $\boldsymbol{V}^*\boldsymbol{V}$, where $\boldsymbol{V}$ is also memoryless, by computing a Cholesky factorization of each diagonal block, i.e. $W_k = V_k^T V_k$, and we denote $\boldsymbol{W}^{1/2} := \boldsymbol{V} = \mathrm{diag}\{V_k\}$; similarly, a sequence of operators $\boldsymbol{W}_n \in \mathcal{M}$ converges to $\boldsymbol{W} \in \mathcal{M}$ in norm if the diagonal blocks of $\boldsymbol{W}_n$ converge uniformly to the diagonal blocks of $\boldsymbol{W}$ in norm.

Note that in this section we have talked almost exclusively of bounded operators. The framework introduced here can also be used to deal with unbounded operators. This is done by considering an *extended* space $\mathcal{B}_e$ which consists of the set of operators for which the projection $\boldsymbol{P}_n\boldsymbol{G} \in \mathcal{B}$ for all $n \in \mathbb{Z}$; see for example [29]. Note that the extended space $\mathcal{B}_e$ is the completion of $\mathcal{B}$ in the *resolution* topology; eg. [35].

Finally we introduce the notation used for closed-loop systems. The open-loop system

$$\boldsymbol{G} = \begin{bmatrix} \boldsymbol{G}_{11} & \boldsymbol{G}_{12} \\ \boldsymbol{G}_{21} & \boldsymbol{G}_{22} \end{bmatrix}$$

represents the mapping

$$\begin{aligned} \boldsymbol{z} &= \boldsymbol{G}_{11}\boldsymbol{w} + \boldsymbol{G}_{12}\boldsymbol{u} \\ \boldsymbol{y} &= \boldsymbol{G}_{21}\boldsymbol{w} + \boldsymbol{G}_{22}\boldsymbol{u} \end{aligned}$$

With a controller $\boldsymbol{K}$, the feedback $\boldsymbol{u} = \boldsymbol{K}\boldsymbol{y}$ results in the closed-loop system given by the linear fractional transformation

$$\mathcal{F}_\ell(\boldsymbol{G}, \boldsymbol{K}) := \boldsymbol{G}_{11} + \boldsymbol{G}_{12}\boldsymbol{K}(\boldsymbol{I} - \boldsymbol{G}_{22}\boldsymbol{K})^{-1}\boldsymbol{G}_{21}$$

whenever the indicated inverse exists.

In the next section we will consider a special class of operators, namely those that can be represented by a state-space realization.

2.2 State-space realizations

Consider a causal operator $\boldsymbol{G}$ admitting a state-space realization

$$\Sigma_G := \left\{ \begin{array}{rcl} \boldsymbol{Z}\boldsymbol{x} & = & \boldsymbol{A}\boldsymbol{x} + \boldsymbol{B}\boldsymbol{w} \\ \boldsymbol{z} & = & \boldsymbol{C}\boldsymbol{x} + \boldsymbol{D}\boldsymbol{w} \end{array} \right. =: \left[\begin{array}{c|c} \boldsymbol{A} & \boldsymbol{B} \\ \hline \boldsymbol{C} & \boldsymbol{D} \end{array} \right] \tag{2.1}$$

Here the operator $\boldsymbol{A} = \text{diag}\{A_k\} \in \mathcal{M}(\ell_2^n, \ell_2^n)$; with similar expressions for $\boldsymbol{B} \in \mathcal{M}(\ell_2^m, \ell_2^n)$, $\boldsymbol{C} \in \mathcal{M}(\ell_2^n, \ell_2^p)$ and $\boldsymbol{D} \in \mathcal{M}(\ell_2^m, \ell_2^p)$. As we have done above, we will usually assume that these sequences are all bounded. The operator $\boldsymbol{G}$, mapping $\boldsymbol{w} \in \ell_2^m$ to $\boldsymbol{z} \in \ell_2^p$, will often be called the system $\boldsymbol{G}$.

We say that the operator $\boldsymbol{A}$ is uniformly exponentially stable (UES) if there exist constants $c > 0$ and $\beta \in [0, 1)$ such that for all $k \in \mathbb{Z}$ and $l \in \mathbb{N}$

$$\|A_{k+l-1}A_{k+l-2}\dots A_k\| \leq c\beta^l. \tag{2.2}$$

If $\boldsymbol{A}$ is UES, we say that the corresponding system (2.1) is stable.

By defining the spectral radius of an operator $\boldsymbol{T} \in \mathcal{B}(\ell_2^n, \ell_2^n)$ as

$$\rho(\boldsymbol{T}) := \lim_{n\to\infty} \|\boldsymbol{T}^n\|^{1/n}$$

the notion of uniform exponential stability can be characterized in a way which is reminiscent of the well known result for LTI systems.

Lemma 2.5 *Suppose that the operator* $\boldsymbol{A}$ *is in* $\mathcal{M}(\ell_2^n, \ell_2^n)$*. Then* $\boldsymbol{A}$ *is UES if and only if* $\rho(\boldsymbol{Z}^*\boldsymbol{A}) < 1$. □

Proof: This result was proved in [59, Theorem 4.5] in reference to uniform asymptotic stability. We provide a self-contained proof for completeness.
(Necessity) Consider

$$\nu_l := \|(\boldsymbol{Z}^*\boldsymbol{A})^l\| = \|(\boldsymbol{Z}^*\boldsymbol{A})^l\boldsymbol{Z}^l\| = \sup_k \|A_{k+l-1}\cdots A_k\| \leq c\beta^l$$

It follows that

$$\rho(\boldsymbol{Z}^*\boldsymbol{A}) = \lim_{l\to\infty} {\nu_l}^{1/l} \leq \beta \lim_{l\to\infty} c^{1/l} = \beta < 1$$

(Sufficiency) Since $\lim_{l\to\infty} {\nu_l}^{1/l} < 1$ we know that for some $\epsilon > 0$, there exists $L \in \mathbb{N}$ such that for all $l > L$

$${\nu_l}^{1/l} < 1 - \epsilon$$

Let $\beta := (1-\epsilon)$ and $c := \max\{\max_{1\le l\le L} \nu_l, 1\}/\beta^L$. It follows that for $l \le L$

$$\begin{aligned}\|A_{k+l-1}\cdots A_k\| &\le \sup_k \|A_{k+l-1}\cdots A_k\| \\ &= \nu_l \le \max_{1\le l\le L} \nu_l \le c\beta^L \le c\beta^l\end{aligned}$$

Similarly, for $l > L$

$$\begin{aligned}\|A_{k+l-1}\cdots A_k\| &\le \sup_k \|A_{k+l-1}\cdots A_k\| \\ &= \nu_l < (1-\epsilon)^l = \beta^l \le c\beta^l\end{aligned}$$

which completes the proof. ■

Remark 2.6 For an LTI system where all the diagonal elements of $\boldsymbol{A}$ are A we have that $(\boldsymbol{Z}^*\boldsymbol{A})^k = (\boldsymbol{Z}^*)^k\boldsymbol{A}^k$. Since $\boldsymbol{Z}^*$ is unitary, we have

$$\|(\boldsymbol{Z}^*)^k\boldsymbol{A}^k\| = \|\boldsymbol{A}^k\|$$

and hence $\rho(\boldsymbol{Z}^*\boldsymbol{A}) = \rho(A)$. Thus, this lemma says that the LTI system is UES iff all of the eigenvalues of A have norm less than 1.

Remark 2.7 The operator $\boldsymbol{Z}^*\boldsymbol{A}$ is a *weighted shift*. Many properties associated with these types of operators, including computation of the spectral radius, are known [74].

From this result it is easy to see that if $\boldsymbol{A}$ is UES, then $(\boldsymbol{I}-\boldsymbol{Z}^*\boldsymbol{A})$ is invertible in $\mathcal{C}(\ell_2^n, \ell_2^n)$, and that

$$\Phi := (\boldsymbol{I}-\boldsymbol{Z}^*\boldsymbol{A})^{-1} = \sum_{k=0}^{\infty}(\boldsymbol{Z}^*\boldsymbol{A})^k \tag{2.3}$$

where the right-hand side converges in norm. Hence, if $\boldsymbol{A}$ is UES, the state-space realization (2.1) results in the causal (bounded) operator

$$\boldsymbol{G} := \boldsymbol{C}\Phi\boldsymbol{Z}^*\boldsymbol{B} + \boldsymbol{D}$$

mapping $\boldsymbol{w}$ to $\boldsymbol{z}$. It is easy to see that the elements of $\boldsymbol{G}$ can be written as

$$G_{k,l} = \begin{cases} D_k & \text{for } k = l \\ C_k B_{k-1} & \text{for } k = l+1 \\ C_k A_{k-1}\cdots A_{l+1}B_l & \text{for } k > l+1 \\ 0 & \text{elsewhere}\end{cases}$$

As a result from Lemma 2.5, we note the following:

Lemma 2.8 *Suppose that* $\boldsymbol{W}, \boldsymbol{A} \in \mathcal{M}(\ell_2^n, \ell_2^n)$, $\boldsymbol{W} \geq \boldsymbol{0}$ *and that* $\boldsymbol{A}$ *is UES. Then the operator Stein equation*

$$\boldsymbol{L} = \boldsymbol{A}^*\boldsymbol{Z}\boldsymbol{L}\boldsymbol{Z}^*\boldsymbol{A} + \boldsymbol{W} \tag{2.4}$$

has a unique solution $\boldsymbol{L} \in \mathcal{M}(\ell_2^n, \ell_2^n)$ *and* $\boldsymbol{L} \geq \boldsymbol{0}$. □

Proof: As the necessity part of the proof is fairly lengthy, we refer to [63]. We will give the sufficiency proof, since it shows the use of Lemma 2.5. With $\boldsymbol{L}_0 := \boldsymbol{W}$ and for $k > 0$, define the sequence

$$\boldsymbol{L}_k := \boldsymbol{A}^*\boldsymbol{Z}\boldsymbol{L}_{k-1}\boldsymbol{Z}^*\boldsymbol{A} + \boldsymbol{W} = \sum_{j=0}^{k} (\boldsymbol{A}^*\boldsymbol{Z})^j \boldsymbol{W} (\boldsymbol{Z}^*\boldsymbol{A})^j$$

Now,

$$\|(\boldsymbol{A}^*\boldsymbol{Z})^j \boldsymbol{W} (\boldsymbol{Z}^*\boldsymbol{A})^j\| \leq \|\boldsymbol{W}\| \|(\boldsymbol{Z}^*\boldsymbol{A})^j\|^2$$

Since $\boldsymbol{A}$ is stable, from Lemma 2.5 we know that there exists $N \in \mathbb{N}$ and some ϵ such that $\|(\boldsymbol{Z}^*\boldsymbol{A})^n\| \leq 1 - \epsilon$, for all $n \geq N$. Thus

$$n \geq N \Rightarrow \|(\boldsymbol{A}^*\boldsymbol{Z})^n \boldsymbol{W} (\boldsymbol{Z}^*\boldsymbol{A})^n\| \leq \|\boldsymbol{W}\|(1-\epsilon)^2$$

It follows that the sequence $\boldsymbol{L}_k$ converges to the solution to (2.4). Uniqueness also follows easily. ■

The dual of this will also be used.

Corollary 2.9 *Suppose that* $\boldsymbol{W}, \boldsymbol{A} \in \mathcal{M}(\ell_2^n, \ell_2^n)$, $\boldsymbol{W} \geq \boldsymbol{0}$ *and that* $\boldsymbol{A}$ *is UES. Then the operator Stein equation*

$$\bar{\boldsymbol{L}} = \boldsymbol{A}\boldsymbol{Z}^*\bar{\boldsymbol{L}}\boldsymbol{Z}\boldsymbol{A}^* + \boldsymbol{W}$$

has a unique solution $\bar{\boldsymbol{L}} \in \mathcal{M}(\ell_2^n, \ell_2^n)$ *and* $\bar{\boldsymbol{L}} \geq \boldsymbol{0}$. □

For systems, we say that the pair $(\boldsymbol{A}, \boldsymbol{B})$ is uniformly stabilizable (respectively $(\boldsymbol{C}, \boldsymbol{A})$ is uniformly detectable) if there exists a bounded memoryless operator $\boldsymbol{F}$ (resp. $\boldsymbol{H}$) such that $\boldsymbol{A} + \boldsymbol{B}\boldsymbol{F}$ (resp. $\boldsymbol{A} + \boldsymbol{H}\boldsymbol{C}$) is UES. The following result will be useful. It connects the notions of input-output stability (i.e. boundedness of the operator) and internal stability (UES of the operator $\boldsymbol{A}$.)

Lemma 2.10 ([2]) *Suppose that for a system (2.1) the pairs* $(\boldsymbol{A}, \boldsymbol{B})$ *and* $(\boldsymbol{C}, \boldsymbol{A})$ *are uniformly exponentially stabilizable and detectable. Then the corresponding operator* $\boldsymbol{G}$ *is in* $\mathcal{B}(\ell_2^m, \ell_2^p)$ *if and only if* $\boldsymbol{A}$ *is UES.* □

In the next section we will introduce the time-reverse operator, which we will use to treat anti-causal systems.

2.3 Time-reverse systems

The time-reverse operator Ω is defined in [45] as

$$(\Omega \boldsymbol{x})_k = x_{-k}$$

This operator Ω depends on the dimension of the signal that it is working on, but since this is clear from the context we will omit an index for this dimension. A matrix representation of Ω is given by

$$\Omega_{ij} = \begin{cases} I & \text{if } j = -i \\ 0 & \text{if } j \neq -i \end{cases}$$

and it is easily seen that $\Omega = \Omega^*$ and $\Omega^2 = \boldsymbol{I}$. Using this it is immediate that pre- or post-multiplying an operator with Ω does not change the induced operator norm. Also it is straightforward to check that $(\Omega \boldsymbol{G}^* \Omega)_{i,j} = G^T_{-j,-i}$, and since $\|\boldsymbol{G}^*\| = \|\boldsymbol{G}\|$ we have the following:

Corollary 2.11 *Suppose that $\boldsymbol{G}$ is a bounded operator. Then*

(i) $\|\Omega \boldsymbol{G}^* \Omega\| = \|\boldsymbol{G}\|$

(ii) $\Omega \boldsymbol{G}^* \Omega$ is causal if and only if $\boldsymbol{G}$ is causal. □

These results will be used to compare the causal operator $\boldsymbol{G}$ with the anti-causal operator $\boldsymbol{G}^*$. Another important property to consider is the influence of Ω on the stability of an operator.

Lemma 2.12 *$\boldsymbol{A}$ is UES if and only if $\Omega \boldsymbol{A}^* \Omega$ is UES.*

Proof: By definition $\boldsymbol{A}$ is UES if there exist constants $c > 0$ and $\beta \in [0, 1)$ such that for all $k \in \mathbb{Z}$ and $l \in \mathbb{N}$

$$\begin{aligned} &\|A_{k+l-1} A_{k+l-2} \dots A_k\| \leq c\beta^l \\ \Leftrightarrow &\|A_k^T \dots A_{k+l-2}^T A_{k+l-1}^T\| \leq c\beta^l \\ \Leftrightarrow &\|(\Omega \boldsymbol{A}^* \Omega)_{-k} \dots (\Omega \boldsymbol{A}^* \Omega)_{-k-l+2} (\Omega \boldsymbol{A}^* \Omega)_{-k-l+1}\| \leq c\beta^l \end{aligned}$$

which is equivalent to $\Omega \boldsymbol{A}^* \Omega$ being UES. ■

Using this it is easy to show the following lemma, which is proved in [45] for any state-space realization.

Lemma 2.13 *Suppose that $\boldsymbol{G}$ is given by a stable state-space realization (2.1). Then*

$$\Sigma_{\Omega G^* \Omega} = \left[\begin{array}{c|c} \Omega \boldsymbol{A}^* \Omega & \Omega \boldsymbol{C}^* \Omega \\ \hline \Omega \boldsymbol{B}^* \Omega & \Omega \boldsymbol{D}^* \Omega \end{array} \right]$$

Proof: Since A is UES, Lemma 2.12 tells us that $\Omega A^*\Omega$ is also UES. Hence (2.3) is well-defined for both A and $\Omega A^*\Omega$. Thus

$$\begin{aligned}\Omega G^*\Omega &= \Omega\left(B^*Z(I - A^*Z)^{-1}C^* + D^*\right)\Omega \\ &= \Omega\left(B^*(I - ZA^*)^{-1}ZC^* + D^*\right)\Omega \\ &= \Omega B^*\Omega(I - \Omega ZA^*\Omega)^{-1}\Omega ZC^*\Omega + \Omega D^*\Omega\end{aligned}$$

where we used that $\Omega^2 = I$. It is straightforward to check that $\Omega Z = Z^*\Omega$, from which the result follows. ∎

Finally we will consider the impact of Ω on a closed-loop system. For an open-loop system

$$G = \begin{bmatrix} G_{11} & G_{12} \\ G_{21} & G_{22} \end{bmatrix}$$

we denote

$$\Omega G\Omega = \begin{bmatrix} \Omega G_{11}\Omega & \Omega G_{12}\Omega \\ \Omega G_{21}\Omega & \Omega G_{22}\Omega \end{bmatrix} \tag{2.5}$$

where all the Ω's have appropriate dimensions. For a linear fractional transformation this allows for an easy comparison.

Lemma 2.14 *We have* $\mathcal{F}_\ell(\Omega G^*\Omega, \Omega K^*\Omega) = \Omega\mathcal{F}_\ell(G, K)^*\Omega$.

Proof: This follows directly from (2.5) since

$$\begin{aligned}&\mathcal{F}_\ell(\Omega G^*\Omega, \Omega K^*\Omega) \\ &\quad = \Omega G_{11}^*\Omega + \Omega G_{21}^*\Omega\Omega K^*\Omega(I - \Omega G_{22}^*\Omega\Omega K^*\Omega)^{-1}\Omega G_{12}^*\Omega \\ &\quad = \Omega G_{11}^*\Omega + \Omega G_{21}^*K^*(I - G_{22}^*K^*)^{-1}G_{12}^*\Omega \\ &\quad = \Omega G_{11}^*\Omega + \Omega G_{21}^*(I - K^*G_{22}^*)^{-1}K^*G_{12}^*\Omega \\ &\quad = \Omega\mathcal{F}_\ell(G, K)^*\Omega\end{aligned}$$

where we used again that $\Omega^2 = I$. ∎

From this it follows immediately that $\|\mathcal{F}_\ell(G, K)\| < 1$ if and only if $\|\mathcal{F}_\ell(\Omega G^*\Omega, \Omega K^*\Omega)\| < 1$.

3 Induced Operator Norms

State-space characterizations of the $\mathcal{H}_\infty$ norm for linear time-invariant systems have been obtained for both continuous-time [16] and discrete-time systems [52], by considering the connection between the norm bound $\|G\|_\infty < \gamma$ and the existence of spectral factors of the Hermitian function $I - \gamma^{-2}G^{\sim}G$. In this chapter we provide the analogous characterization for the induced norm of both discrete-time and hybrid linear time-varying systems in terms of operator algebraic Riccati equations.

The approach that we use in our proofs is based on standard optimal control methods along the lines of [75, 76]. What facilitates this approach is an operator theoretic description of discrete-time time-varying systems, introduced in Chapter 2. We show that an operator $\boldsymbol{G}$ has induced norm less than γ if and only if a related operator algebraic Riccati equation has a positive, semi-definite solution which is exponentially stabilizing. These results resemble the corresponding LTI results closely. The characterizations will give us an equivalent expression for the entropy as defined in Chapter 4.

Having obtained the characterization for discrete-time systems, we consider the norm properties of system of the form $G\boldsymbol{H}_h$, where $\boldsymbol{H}_h$ is the hold operator, and $\boldsymbol{S}_h G$, where $\boldsymbol{S}_h$ is the sample operator. We show that in both cases, the norm of these two systems can be written in terms of the norm of a related self-adjoint discrete-time operator.

3.1 Characterizations of the induced norm

In this section we provide characterizations of the induced operator norm associated with the causal system $\boldsymbol{G}$, represented by a state-space representation. The operator theoretic treatment of this problem allows us to consider infinite-horizon systems with relative ease. Of course, much more care is needed in the LTV case since the operators considered are non-compact. Recall the realization (2.1) of $\boldsymbol{G}$ mapping ℓ_2^m to ℓ_2^p:

$$\Sigma_G := \begin{cases} \boldsymbol{Z}\boldsymbol{x} &= \boldsymbol{A}\boldsymbol{x} + \boldsymbol{B}\boldsymbol{w} \\ \boldsymbol{z} &= \boldsymbol{C}\boldsymbol{x} + \boldsymbol{D}\boldsymbol{w} \end{cases}$$

The operators $\boldsymbol{A}$, $\boldsymbol{B}$, $\boldsymbol{C}$, $\boldsymbol{D}$ are all memoryless and bounded. Throughout the section it will be assumed that $\boldsymbol{A}$ is UES, so $\boldsymbol{G}$ is stable (see Section 2.2). Furthermore we will assume γ to be equal to 1. The results for $\gamma \neq 1$ can be obtained by scaling the state-space operators to $\boldsymbol{A}$, $\gamma^{-1/2}\boldsymbol{B}$, $\gamma^{-1/2}\boldsymbol{C}$, and $\gamma^{-1}\boldsymbol{D}$.

Our first result characterizes the norm bound of $\boldsymbol{G}$ in terms of an operator Riccati equation, and shows the equivalence with the existence of a spectral factorization.

Theorem 3.1 *Suppose that $\boldsymbol{G} \in \mathcal{C}$ admits a stable state-space realization. The following statements are equivalent:*

1. $\boldsymbol{I} - \boldsymbol{G}^*\boldsymbol{G} > \boldsymbol{0}$

2. *There exists $\boldsymbol{0} \leq \boldsymbol{X} \in \mathcal{M}$ satisfying the operator algebraic Riccati equation*

$$\begin{aligned} \boldsymbol{X} = {} & \boldsymbol{A}^*\boldsymbol{Z}\boldsymbol{X}\boldsymbol{Z}^*\boldsymbol{A} + \boldsymbol{C}^*\boldsymbol{C} \\ & + (\boldsymbol{A}^*\boldsymbol{Z}\boldsymbol{X}\boldsymbol{Z}^*\boldsymbol{B} + \boldsymbol{C}^*\boldsymbol{D})\,\boldsymbol{V}^{-1}\,(\boldsymbol{B}^*\boldsymbol{Z}\boldsymbol{X}\boldsymbol{Z}^*\boldsymbol{A} + \boldsymbol{D}^*\boldsymbol{C}) \end{aligned} \tag{3.1}$$

   with $\boldsymbol{V} > \boldsymbol{0}$ such that $\boldsymbol{A}_F$ is UES. Here

$$\begin{aligned} \boldsymbol{V} &:= \boldsymbol{I} - \boldsymbol{D}^*\boldsymbol{D} - \boldsymbol{B}^*\boldsymbol{Z}\boldsymbol{X}\boldsymbol{Z}^*\boldsymbol{B} \\ \boldsymbol{A}_F &:= \boldsymbol{A} + \boldsymbol{B}\boldsymbol{V}^{-1}\,(\boldsymbol{B}^*\boldsymbol{Z}\boldsymbol{X}\boldsymbol{Z}^*\boldsymbol{A} + \boldsymbol{D}^*\boldsymbol{C}) \end{aligned}$$

3. *There exists a spectral factorization $\boldsymbol{I} - \boldsymbol{G}^*\boldsymbol{G} = \boldsymbol{M}^*\boldsymbol{M}$ for some operator $\boldsymbol{M} \in \mathcal{C} \cap \mathcal{C}^{-1}$.*

Proof: To prove (3)⇒(1) we decompose $\boldsymbol{M} = \boldsymbol{M}_m + \boldsymbol{M}_{sc}$, where $\boldsymbol{M}_m$ is memoryless and $\boldsymbol{M}_{sc}$ is strictly causal. Since $\boldsymbol{M} \in \mathcal{C} \cap \mathcal{C}^{-1}$, it follows from Lemma 2.3 that $\boldsymbol{M}_m^* \boldsymbol{M}_m > \boldsymbol{0}$, say $\langle \boldsymbol{M}_m \boldsymbol{x}, \boldsymbol{M}_m \boldsymbol{x} \rangle \geq \epsilon \langle \boldsymbol{x}, \boldsymbol{x} \rangle$ for all $\boldsymbol{x} \in \ell_2^m$. Then

$$\begin{aligned}
\langle \boldsymbol{x}, \boldsymbol{x} \rangle - \langle \boldsymbol{G}\boldsymbol{x}, \boldsymbol{G}\boldsymbol{x} \rangle &= \langle \boldsymbol{M}\boldsymbol{x}, \boldsymbol{M}\boldsymbol{x} \rangle \\
&= \langle \boldsymbol{M}_m(\boldsymbol{I} + \boldsymbol{M}_m^{-1}\boldsymbol{M}_{sc})\boldsymbol{x}, \boldsymbol{M}_m(\boldsymbol{I} + \boldsymbol{M}_m^{-1}\boldsymbol{M}_{sc})\boldsymbol{x} \rangle \\
&\geq \epsilon \|(\boldsymbol{I} + \boldsymbol{M}_m \boldsymbol{M}_{sc})^{-1}\|^{-2} \langle \boldsymbol{x}, \boldsymbol{x} \rangle
\end{aligned}$$

for all $\boldsymbol{x} \in \ell_2^m$. To show (2)⇒(3) it is straightforward to see that, if $\boldsymbol{X}$ satisfies (3.1), then a spectral factorization of $\boldsymbol{I} - \boldsymbol{G}^*\boldsymbol{G}$ is given by

$$\boldsymbol{M} = \boldsymbol{V}^{1/2} - \boldsymbol{V}^{-1/2}(\boldsymbol{B}^*\boldsymbol{Z}\boldsymbol{X}\boldsymbol{Z}^*\boldsymbol{A} + \boldsymbol{D}^*\boldsymbol{C})(\boldsymbol{I} - \boldsymbol{Z}^*\boldsymbol{A})^{-1}\boldsymbol{Z}^*\boldsymbol{B}$$

Furthermore

$$\boldsymbol{M}^{-1} = \boldsymbol{V}^{-1/2} + \boldsymbol{V}^{-1}(\boldsymbol{B}^*\boldsymbol{Z}\boldsymbol{X}\boldsymbol{Z}^*\boldsymbol{A} + \boldsymbol{D}^*\boldsymbol{C})(\boldsymbol{I} - \boldsymbol{Z}^*\boldsymbol{A}_F)^{-1}\boldsymbol{Z}^*\boldsymbol{B}\boldsymbol{V}^{-1/2}$$

which is causal, since $\boldsymbol{A}_F$ is UES and all operators are causal [3]. It remains to show that (1)⇒(2). This is a special case of Theorem 6.5, namely with $\boldsymbol{B}_2 = \boldsymbol{D}_{12} = \boldsymbol{0}$. Theorem 6.5 is proven in Appendix A, a direct proof is given in [55]. ■

Since we will show in Chapter 4 that the entropy of an operator $\boldsymbol{G}$, given by a state-space realization, can be written in terms of $\boldsymbol{X}$, it is important that $\boldsymbol{X}$ be unique.

Lemma 3.2 *The Riccati operator equation (3.1) has a unique stabilizing solution.*

Proof: Suppose $\bar{\boldsymbol{X}}$ is also a stabilizing solution of (3.1). Then it is an easy calculation to see that

$$\boldsymbol{X} - \bar{\boldsymbol{X}} = \boldsymbol{A}_{cl\bar{X}}^* \boldsymbol{Z}(\boldsymbol{X} - \bar{\boldsymbol{X}})\boldsymbol{Z}^*\boldsymbol{A}_{clX}$$

where $\boldsymbol{A}_{clX}$ and $\boldsymbol{A}_{cl\bar{X}}$ are the closed-loop matrices corresponding to $\boldsymbol{X}$ and $\bar{\boldsymbol{X}}$ respectively. Since both $\boldsymbol{A}_{clX}$ and $\boldsymbol{A}_{cl\bar{X}}$ are UES, it follows that $\bar{\boldsymbol{X}} = \boldsymbol{X}$. ■

We will also give the dual result of Theorem 3.1, which we will use later for hybrid systems. For this consider the anti-causal system $\boldsymbol{H}$ given by $\boldsymbol{H} = \boldsymbol{G}^*$.

Theorem 3.3 *Suppose that $\boldsymbol{H}^* \in \mathcal{C}$ admits a stable state-space realization. The following statements are equivalent:*

1. $\boldsymbol{I} - \boldsymbol{H}^*\boldsymbol{H} > \boldsymbol{0}$

2. *There exists $\boldsymbol{0} \leq \boldsymbol{Y} \in \mathcal{M}$ satisfying the operator algebraic Riccati equation*

$$\begin{aligned} \boldsymbol{Y} = \boldsymbol{A}\boldsymbol{Z}^*\boldsymbol{Y}\boldsymbol{Z}\boldsymbol{A}^* + \boldsymbol{B}\boldsymbol{B}^* & \\ + \left(\boldsymbol{A}\boldsymbol{Z}^*\boldsymbol{Y}\boldsymbol{Z}\boldsymbol{C}^* + \boldsymbol{B}\boldsymbol{D}^*\right)\boldsymbol{U}^{-1}\left(\boldsymbol{C}\boldsymbol{Z}^*\boldsymbol{Y}\boldsymbol{Z}\boldsymbol{A}^* + \boldsymbol{D}\boldsymbol{B}^*\right) & \end{aligned} \tag{3.2}$$

   with $\boldsymbol{U} > \boldsymbol{0}$ such that $\boldsymbol{A}_L$ is UES. Here

$$\begin{aligned} \boldsymbol{U} &:= \boldsymbol{I} - \boldsymbol{D}\boldsymbol{D}^* - \boldsymbol{C}\boldsymbol{Z}^*\boldsymbol{Y}\boldsymbol{Z}\boldsymbol{C}^* \\ \boldsymbol{A}_L &:= \boldsymbol{A} + (\boldsymbol{A}\boldsymbol{Z}^*\boldsymbol{Y}\boldsymbol{Z}\boldsymbol{C}^* + \boldsymbol{B}\boldsymbol{D}^*)\boldsymbol{U}^{-1}\boldsymbol{C} \end{aligned}$$

3. *There exists a co-spectral factorization $\boldsymbol{I} - \boldsymbol{H}^*\boldsymbol{H} = \boldsymbol{N}\boldsymbol{N}^*$ for some operator $\boldsymbol{N} \in \mathcal{C} \cap \mathcal{C}^{-1}$.*

Proof: (2)⇒(3) and (3)⇒(1) are similar to Theorem 3.1. To show (1)⇒(2) we use the time-reverse operator Ω introduced in Section 2.3. From Corollary 2.11 we know that $\|\boldsymbol{H}^*\| = \|\boldsymbol{H}\| < 1$ if and only if $\|\Omega\boldsymbol{H}\Omega\| < 1$. The operator $\Omega\boldsymbol{H}\Omega$ is causal, and as in Lemma 2.13 we get the state-space realizations:

$$\Sigma_{H^*} = \left[\begin{array}{c|c} \boldsymbol{A} & \boldsymbol{B} \\ \hline \boldsymbol{C} & \boldsymbol{D} \end{array}\right] \quad \text{and} \quad \Sigma_{\Omega H \Omega} = \left[\begin{array}{c|c} \Omega\boldsymbol{A}^*\Omega & \Omega\boldsymbol{C}^*\Omega \\ \hline \Omega\boldsymbol{B}^*\Omega & \Omega\boldsymbol{D}^*\Omega \end{array}\right]$$

From Lemma 2.12 we know that $\Omega\boldsymbol{A}^*\Omega$ is UES since $\boldsymbol{A}$ is. Applying Theorem 3.1 to $\Omega\boldsymbol{H}\Omega$ gives

$$\begin{aligned} \boldsymbol{X} = \Omega\boldsymbol{A}\Omega\boldsymbol{Z}\boldsymbol{X}\boldsymbol{Z}^*\Omega\boldsymbol{A}\Omega + \Omega\boldsymbol{B}\boldsymbol{B}^*\Omega & \\ + \Omega(\boldsymbol{A}\Omega\boldsymbol{Z}\boldsymbol{X}\boldsymbol{Z}^*\Omega\boldsymbol{C}^* + \boldsymbol{B}\boldsymbol{D}^*)\boldsymbol{V}^{-1}(\boldsymbol{C}\Omega\boldsymbol{Z}\boldsymbol{X}\boldsymbol{Z}^*\Omega\boldsymbol{A}^* + \boldsymbol{D}\boldsymbol{B}^*)\Omega & \end{aligned}$$

where $\boldsymbol{V} > \boldsymbol{0}$ and $\boldsymbol{A}_F$ is UES. Here

$$\begin{aligned} \boldsymbol{V} &= \boldsymbol{I} - \boldsymbol{D}\boldsymbol{D}^* - \boldsymbol{C}\Omega\boldsymbol{Z}\boldsymbol{X}\boldsymbol{Z}^*\Omega\boldsymbol{C}^* \\ \boldsymbol{A}_F &= \Omega\boldsymbol{A}^*\Omega + \Omega\boldsymbol{C}^*\boldsymbol{V}^{-1}(\boldsymbol{C}\Omega\boldsymbol{Z}\boldsymbol{X}\boldsymbol{Z}^*\Omega\boldsymbol{A}^* + \boldsymbol{D}\boldsymbol{B}^*)\Omega \end{aligned}$$

Notice that we used repeatedly that $\Omega^2 = \boldsymbol{I}$. Since $\Omega\boldsymbol{Z}\Omega = \boldsymbol{Z}^*$, the operator Riccati equation (3.2) follows by pre- and post-multiplying with Ω and setting $\boldsymbol{Y} = \Omega\boldsymbol{X}\Omega$. The stability result follows from Lemma 2.12, since $\boldsymbol{A}_L = \Omega\boldsymbol{A}_F^*\Omega$. ■

Remark 3.4 The characterizations of the induced norm presented here, together with the accompanying spectral factors, are instrumental in considering minimum entropy control problems for time-varying systems. In Chapter 4, the entropy of a discrete-time system is defined in terms of the memoryless part of the spectral factor defined in Theorem 3.1.

Remark 3.5 The equivalence between the norm bound and the existence of stabilizing solutions to the operator Riccati equations in Theorems 3.1 and 3.3 is usually known as the bounded real lemma. The connection between this result, and the related positive real lemma of network synthesis is outlined in the monograph [4]. For discrete-time systems, the positive real lemma was first derived in [48]. The connection with the $\mathcal{H}_\infty$ norm of the transfer function seems to have been first made by Willems [81].

In the next section we will find results similar to Theorem 3.1 and Theorem 3.3 to characterize the induced operator norm of hybrid systems.

3.2 Time-varying hybrid systems

Hybrid systems are closed-loop systems consisting of a continuous-time plant controlled by a discrete-time controller that is interfaced to the plant via ideal sample-and-hold circuitry [21]. A necessary step for characterizing the induced norm of hybrid feedback systems is a characterization of the induced norms of systems of the form $G\boldsymbol{H}_h$ and $\boldsymbol{S}_hG$. Here G is a continuous-time plant, and $\boldsymbol{S}_h$ and $\boldsymbol{H}_h$ are the ideal sampler and hold operators respectively, where h is the sampling (respectively holding) time.

Until only recently, research on hybrid systems had been restricted to LTI systems. Feedback stability results of Chen and Francis have been extended to hybrid systems comprising of linear time-varying components in [50]. Motivation for this research can be found in the fact that optimal controllers based on the $\mathcal{H}_\infty$ control paradigm are now being introduced to practical systems which are time-varying.

In [19, 46] it is shown that, in the LTI case, the norm of $G\boldsymbol{H}_h$ and $\boldsymbol{S}_hG$ can be obtained by computing the norm of a related discrete-time LTI system. This is done by using a fact, first used in [19], that although $G\boldsymbol{H}_h$ (resp. $\boldsymbol{S}_hG$) maps $\ell_2 \to \mathcal{L}_2$ (resp. $\mathcal{L}_2 \to \ell_2$) one may instead consider the operator $(G\boldsymbol{H}_h)^*G\boldsymbol{H}_h$ (resp. $\boldsymbol{S}_hG(\boldsymbol{S}_hG)^*$) which maps ℓ_2 to ℓ_2. Since the latter operators are both mapping discrete-time signals to discrete-time

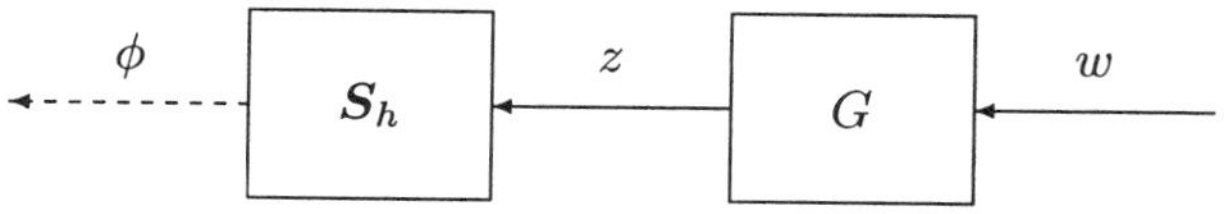

Figure 3.1: Sampled system

signals, results similar to those from the first section of this chapter may be used to characterize the norm of both $G\boldsymbol{H}_h$ and $\boldsymbol{S}_h G$.

A complication that arises in the study of LTV systems which is not present in the corresponding results for LTI systems, is that the operator that is obtained from taking $(G\boldsymbol{H}_h)^* G\boldsymbol{H}_h$ or $\boldsymbol{S}_h G(\boldsymbol{S}_h G)^*$ is neither causal nor anti-causal. This requires special care, which we will do by exploiting some recent results of Van der Veen and Verhaegen on spectral factorizations of positive operators [78].

3.2.1 Sampled continuous-time systems

In the class of hybrid systems we first look at sampled continuous-time time-varying systems. Consider the linear continuous-time, time-varying system with state-space representation:

$$\begin{aligned} \dot{x}_t &= A_t x_t + B_t w_t \\ z_t &= C_t x_t \end{aligned} \tag{3.3}$$

defined for $t \in \mathbb{R}$. The direct feedthrough D is taken to be equal to zero in order to avoid the sampling of signals which are not in ℓ_2. The matrices A, B and C are all uniformly bounded functions of t, and A is assumed to be exponentially stable. That is, if $\Phi_{t,\tau}$ denotes the state transition matrix of the homogeneous part of (3.3), there exist positive constants c, β such that

$$\|\Phi_{t,\tau}\| \leq ce^{-\beta(t-\tau)}$$

for all t and τ such that $t \geq \tau$. The impulse response of this system is

$$g_{t,\tau} := \begin{cases} C_t \Phi_{t,\tau} B_\tau & t > \tau \\ 0 & \text{elsewhere} \end{cases} \tag{3.4}$$

The system of Figure 3.1 can be written with a state-space representation

$$\begin{aligned} \dot{x}_t &= A_t x_t + B_t w_t \\ \phi_k &= C_{kh} x_{kh} \end{aligned} \tag{3.5}$$

This gives the expression

$$\phi_k = \int_{-\infty}^{kh} g_{kh,\tau} w_\tau \, d\tau$$

The sampled system, described by (3.5), maps continuous-time signals to discrete-time signals. Hence the computation of the induced norm involves both $\mathcal{L}_2$ as well as ℓ_2 signals. As in [19], where the authors consider the time-invariant case, we can overcome this by relating the induced norm of the sampled system to the induced norm of a discrete-time system. In order to do this we define the controllability matrix by

$$W_{t_i,t_f} := \int_{t_i}^{t_f} \Phi_{t_f,t} B_t B_t^T \Phi_{t_f,t}^T \, dt$$

and the block-diagonal operators

$$\begin{aligned} \boldsymbol{A} &:= \operatorname{diag}\{\Phi_{(k+1)h,kh}\} \\ \boldsymbol{B} &:= \operatorname{diag}\{\Phi_{(k+1)h,kh} W_{-\infty,kh} C_{kh}^T\} \\ \boldsymbol{C} &:= \operatorname{diag}\{C_{kh}\} \\ \boldsymbol{W} &:= \operatorname{diag}\{W_{-\infty,kh}\} \end{aligned}$$

Remark 3.6 Note that $\boldsymbol{A}$ is UES since the discrete-time transition matrix is given by

$$\hat{\Phi}_{k,l} := A_{k-1} A_{k-2} \cdots A_l = \Phi_{kh,lh}$$

for $k > l$. This satisfies

$$\left\|\hat{\Phi}_{k,l}\right\| = \|\Phi_{kh,lh}\| \leq c e^{-\beta([k-l]h)} =: c\hat{\beta}^{k-l}$$

with $\hat{\beta} := \exp(-\beta h)$, $|\hat{\beta}| < 1$, and thus (see (2.2)) $\boldsymbol{A}$ is discrete-time UES.

Our next result characterizes the norm of the sampled system $\boldsymbol{S}_h G$ in terms of the induced operator norm of a discrete-time system.

Theorem 3.7 *The norm of $\boldsymbol{S}_h G : \mathcal{L}_2^m \to \ell_2^p$ equals $\|\boldsymbol{Q}\|^{1/2}$, where $\boldsymbol{Q}$ is the discrete-time operator*

$$\boldsymbol{Q} = \boldsymbol{R} + \boldsymbol{R}^* + \boldsymbol{C}\boldsymbol{W}\boldsymbol{C}^* \tag{3.6}$$

Here $\boldsymbol{R}$ is a causal operator with state-space representation

$$\Sigma_R := \left[\begin{array}{c|c} \boldsymbol{A} & \boldsymbol{B} \\ \hline \boldsymbol{C} & \boldsymbol{0} \end{array}\right]$$

Proof: Define $\boldsymbol{T} := \boldsymbol{S}_h G$. If $\phi = \{\phi_k\} = \boldsymbol{T}w$, then

$$\phi_k = \int_{-\infty}^{kh} g_{kh,\tau} w_\tau \, d\tau$$

We are interested in computing the adjoint $\boldsymbol{T}^* : \ell_2^p \to \mathcal{L}_2^m$. Suppose that $\boldsymbol{v} \in \ell_2^p$ and $r \in \mathcal{L}_2^m$ are such that $r = \boldsymbol{T}^*\boldsymbol{v}$. We will compare the inner products on the Hilbert spaces $\mathcal{L}_2^m$ and ℓ_2^p. First of all

$$\langle \boldsymbol{T}^*\boldsymbol{v}, w\rangle_{\mathcal{L}_2^m} = \int_{-\infty}^{\infty} r_\tau^T w_\tau \, d\tau$$

and thus

$$\begin{aligned}
\langle \boldsymbol{v}, \boldsymbol{T}w\rangle_{\ell_2^p} &= \sum_{k=-\infty}^{\infty} v_k^T \phi_k \\
&= \sum_{k=-\infty}^{\infty} v_k^T \left(\int_{-\infty}^{\infty} g_{kh,\tau} w_\tau \, d\tau \right) \\
&= \int_{-\infty}^{\infty} \left(\sum_{k=-\infty}^{\infty} v_k^T g_{kh,\tau} \right) w_\tau \, d\tau
\end{aligned}$$

Note that $g_{kh,\tau} = 0$ for $\tau \geq kh$. Hence $r(\tau) = \sum_{k=-\infty}^{\infty} g_{kh,\tau}^T v_k$.

Now, we wish to find an operator representation for $\boldsymbol{Q} := \boldsymbol{T}\boldsymbol{T}^* : \ell_2^p \to \ell_2^p$. Suppose that $\phi = \{\phi_k\} = \boldsymbol{T}r$ and $r = \boldsymbol{T}^*\boldsymbol{v}$, then

$$\begin{aligned}
\phi_k &= \int_{-\infty}^{\infty} g_{kh,\tau} r_\tau \, d\tau \\
&= \int_{-\infty}^{\infty} g_{kh,\tau} \left(\sum_{l=-\infty}^{\infty} g_{lh,\tau}^T v_l \right) d\tau \\
&= \sum_{l=-\infty}^{\infty} \left(\int_{-\infty}^{\infty} g_{kh,\tau} g_{lh,\tau}^T \, d\tau \right) v_l
\end{aligned}$$

Substituting from (3.4), we have

$$g_{kh,\tau} g_{lh,\tau}^T = C_{kh} \Phi_{kh,\tau} B_\tau B_\tau^T \Phi_{lh,\tau}^T C_{lh}^T$$

for $kh > \tau$ and $lh > \tau$; zero elsewhere. Thus

$$\begin{aligned} Q_{k,l} &= \int_{-\infty}^{\infty} g_{kh,\tau} g_{lh,\tau}^T \, d\tau \\ &= \int_{-\infty}^{\min(k,l)h} g_{kh,\tau} g_{lh,\tau}^T \, d\tau \\ &= C_{kh} \left(\int_{-\infty}^{\min(k,l)h} \Phi_{kh,\tau} B_\tau B_\tau^T \Phi_{lh,\tau}^T \, d\tau \right) C_{lh}^T \end{aligned}$$

In particular, note that the diagonal elements satisfy

$$\begin{aligned} Q_{k,k} &= C_{kh} \left(\int_{-\infty}^{kh} \Phi_{kh,\tau} B_\tau B_\tau^T \Phi_{kh,\tau}^T \, d\tau \right) C_{kh}^T \\ &= C_{kh} W_{-\infty,kh} C_{kh}^T \end{aligned}$$

and, for $k > l \geq 0$

$$\begin{aligned} Q_{k,l} &= C_{kh} \Phi_{kh,lh} \left(\int_{-\infty}^{lh} \Phi_{lh,\tau} B_\tau B_\tau^T \Phi_{lh,\tau}^T \, d\tau \right) C_{lh}^T \\ &= C_{kh} \Phi_{kh,(l+1)h} \left(\Phi_{(l+1)h,lh} W_{-\infty,lh} C_{lh}^T \right) \\ &=: C_k \hat{\Phi}_{k,l+1} B_l \end{aligned}$$

which is the impulse response of the discrete-time system given by $\boldsymbol{R}$. Moreover, we have that $Q_{k,l} = Q_{l,k}^T$. The operator $\boldsymbol{Q} := \{Q_{k,l}\}$ can therefore be decomposed as

$$\boldsymbol{Q} = \boldsymbol{R} + \boldsymbol{R}^* + \Lambda$$

where $\Lambda := \operatorname{diag}\{Q_{k,k}\} = \boldsymbol{CWC}^*$. It remains to prove that $\boldsymbol{R}$ and $\boldsymbol{Q}$ are bounded; this follows readily from the boundedness of $\boldsymbol{C}$ and $\boldsymbol{W}$ and the UES property of the continuous-time system. It is well-known that $\|\boldsymbol{T}\| = \|\boldsymbol{T}^*\boldsymbol{T}\|^{1/2}$, which completes the proof. ■

Remark 3.8 For finite-dimensional matrices, decomposing them into their upper triangular, diagonal, and lower triangular components is a trivial operation. For bounded, infinite-dimensional operators, this is not the case. In fact, there are operators for which the decomposition (3.6) is not possible as it may lead to unbounded operators. A classical example is the one given in [35, page 56]: the operator $\boldsymbol{Q}$ is the semi-infinite Toeplitz operator

$\boldsymbol{Q} = \boldsymbol{Q_C} + \boldsymbol{Q_A}$, where $\boldsymbol{Q_C}$ is a strictly lower triangular operator which elements are given by

$$Q_{C_{k,l}} := \frac{1}{k-l} \quad \text{for } k > l \geq 0$$

and $\boldsymbol{Q_A} := -\boldsymbol{Q}_C^*$ is strictly anti-causal. A calculation of the induced norms results in $\|\boldsymbol{Q}\| = \pi$, while $\|\boldsymbol{Q_C}\| = \|\boldsymbol{Q_A}\| = \infty$.

In the specific case of operators arising from the sampled-data systems, however, Theorem 3.7 shows that the resulting operators are bounded, as desired.

Unlike the situation that arises in the LTI case, in order to compute the norm of $\boldsymbol{Q}$, and hence the norm of $\boldsymbol{S}_h G$, Theorems 3.1 and 3.3 cannot be used directly. The reason for this is that those theorems apply only to causal or anti-causal operators, but $\boldsymbol{Q}$ is neither. In the LTI case, the corresponding transfer function $Q(z)$ is in ℓ_∞ and so the same problem does not arise. For the LTV systems considered in this paper, we can circumvent this restriction using the following algorithm:

1. Compute $\boldsymbol{Q}$.

2. Find a factorization of $\boldsymbol{Q} = \boldsymbol{PP}^*$, assuming one exists.

3. Find the norm of the causal operator $\boldsymbol{P}$.

We proceed as above and begin by finding a factorization for $\boldsymbol{Q}$. We may then apply Theorem 3.3 to the co-spectral factor $\boldsymbol{P}$. Note that $\boldsymbol{Q} = \boldsymbol{TT}^*$, hence $\boldsymbol{Q} \geq \boldsymbol{0}$. If $\boldsymbol{Q} > \boldsymbol{0}$, then a factorization $\boldsymbol{Q} = \boldsymbol{PP}^*$ is given in [78] by the state-space realization

$$\Sigma_P := \left[\begin{array}{c|c} \boldsymbol{A} & (\boldsymbol{B} - \boldsymbol{AZ^*SZC^*})(\Lambda - \boldsymbol{CZ^*SZC^*})^{-1/2} \\ \hline \boldsymbol{C} & (\Lambda - \boldsymbol{CZ^*SZC^*})^{1/2} \end{array}\right]$$

where $\boldsymbol{0} \leq \boldsymbol{S} \in \mathcal{M}$ is the solution of

$$\begin{aligned} \boldsymbol{S} = {} & \boldsymbol{AZ^*SZA^*} \\ & + (\boldsymbol{AZ^*SZC^*} - \boldsymbol{B})(\Lambda - \boldsymbol{CZ^*SZC^*})^{-1}(\boldsymbol{CZ^*SZA^*} - \boldsymbol{B}^*) \end{aligned} \tag{3.7}$$

such that $\Lambda - \boldsymbol{CZ^*SZC^*} > \boldsymbol{0}$. Now, since $\boldsymbol{P}$ is causal, we can find characterizations for the induced operator norm of $\boldsymbol{Q}$ in terms of the state-space operators describing this factor.

Theorem 3.9 *Suppose that $Q > 0$. Then the following statements are equivalent:*

1. $I - Q > 0$

2. *There exists $0 \leq Y \in \mathcal{M}$ satisfying the operator Riccati equation*

$$Y = AZ^*YZA^* + (AZ^*YZC^* + B)\,U^{-1}\,(CZ^*YZA^* + B^*) \tag{3.8}$$

with $U > 0$ such that A_L is UES. Here

$$U := I - \Lambda - CZ^*YZC^*$$
$$A_L := A + (AZ^*YZC^* + B)U^{-1}C$$

3. *There exists a co-spectral factorization $I - Q = NN^*$ for some operator $N \in \mathcal{C} \cap \mathcal{C}^{-1}$.*

Proof: (3)⇒(1) follows in a similar way as in Theorem 3.1. (2)⇒(3) is also straightforward by defining

$$N = U^{1/2} - CZ^*(I - AZ^*)^{-1}(AZ^*YZC^* + B)U^{-1/2}$$

where Y satisfies (3.8). To show that (1)⇒(2), we first apply Theorem 3.3 to P (note that $\|Q\| = \|P\|^2$) and find that

$$\begin{aligned}\bar{Y} = {} & AZ^*\bar{Y}ZA^* \\ & + (AZ^*SZC^* - B)\,(\Lambda - CZ^*SZC^*)^{-1}\,(CZ^*SZA^* - B^*) \\ & + (AZ^*(\bar{Y} - S)ZC^* + B)V^{-1}(CZ^*(\bar{Y} - S)ZA^* + B^*)\end{aligned} \tag{3.9}$$

has a solution $\bar{Y} \geq 0$ with $\bar{U} > 0$ such that $\bar{A}_F$ is UES. Here

$$\bar{U} := I - \Lambda - CZ^*(\bar{Y} - S)ZC^*$$
$$\bar{A}_L := A + (AZ^*(\bar{Y} - S)ZC^* + B)V^{-1}C$$

By subtracting (3.7) from (3.9) it follows immediately that $Y = \bar{Y} - S$ satisfies (3.8). Also $A_L = \bar{A}_L$ which is UES, and since A is UES it follows from Corollary 2.9 that $Y \geq 0$. ■

Remark 3.10 The theorem is proved under the assumption that $Q > 0$, but is also true in the case where $Q \geq 0$. This can be shown using continuity arguments, while it can also be shown by applying the result of [78] to find a spectral factorization of $I - Q$.

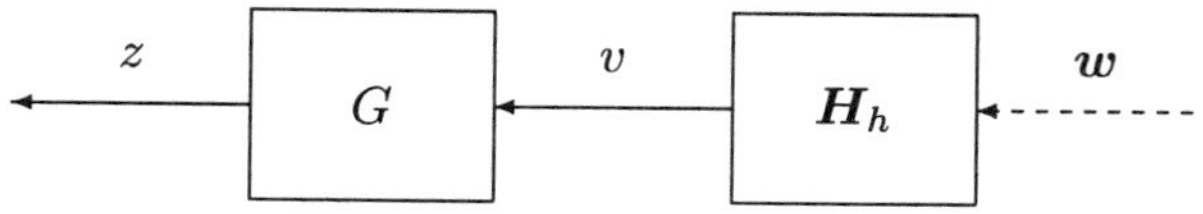

Figure 3.2: System with piecewise constant inputs

An immediate consequence is the following:

Corollary 3.11 *The norm of* $\boldsymbol{H}_h \boldsymbol{S}_h G : \mathcal{L}_2^m \to \mathcal{L}_2^p$ *equals* $\sqrt{h}\,\|\boldsymbol{Q}\|^{1/2}$.

Proof: The result is straightforward by noting that for any sequence in $y \in \ell_2^p$, the $\mathcal{L}_2^p$ norm of $H_h y$ equals $\sqrt{h}\,\|\boldsymbol{y}\|_2$. ■

Remark 3.12 It was shown in [20] that for LTI systems, taking the limit as $h \downarrow 0$ will guarantee that $\|G - \boldsymbol{H}_h \boldsymbol{S}_h G\| = 0$. This, however, is not the case for LTV systems unless the function C_t satisfies an additional slowly time-varying assumption [50]. See also Lemma 7.12.

3.2.2 Continuous-time systems with piecewise constant inputs

In this section we compute the induced norm of an operator mapping discrete-time signals to continuous-time signals. Let G be a continuous-time, time-varying system with state-space representation

$$\begin{aligned} \dot{x}_t &= A_t x_t + B_t w_t \\ z_t &= C_t x_t + D_t w_t \end{aligned} \tag{3.10}$$

defined for $t \in \mathbb{R}$. The same assumptions as in (3.3) concerning boundedness and stability will be made. Note, however, that we can allow a direct feedthrough term D. To the input signal w we first apply the hold operator $\boldsymbol{H}_h$, as shown in Figure 3.2. The state-space representation of the resulting system is given by

$$\begin{aligned} \dot{x}_t &= A_t x_t + B_t v_t \\ z_t &= C_t x_t + D_t v_t \\ v_t &= w_{kh}, \qquad\qquad t \in [kh, (k+1)h) \end{aligned}$$

For $t \in [kh, (k+1)h)$ this can be written as

$$x_t = x_{kh} + \int_{kh}^{t} \Phi_{t,\tau} B_\tau \, d\tau \; w_{kh}$$
$$z_t = C_t x_t + D_t w_{kh}$$

To characterize the induced operator norm of the system where the hold operator is applied to the input, we will proceed in a manner similar to the characterization of the norm on $\boldsymbol{S}_h G$. That is, we first find an expression for $\boldsymbol{T}^*\boldsymbol{T}$, where $\boldsymbol{T} = G\boldsymbol{H}_h$. We then compute two factorizations. We first write $\boldsymbol{Q} := \boldsymbol{T}^*\boldsymbol{T} = \Lambda + \boldsymbol{R} + \boldsymbol{R}^*$ where $\Lambda \in \mathcal{M}$, $\boldsymbol{R} \in \mathcal{C}$, and then use this factorization to compute a spectral factorization of $\boldsymbol{I} - \boldsymbol{Q}$. Before stating our result, recall that the observability Gramian of a pair (C, A) defined on an interval $[t_i, t_f]$ is given by

$$M_{t_i,t_f} = \int_{t_i}^{t_f} \Phi_{t,t_i}^T C_t^T C_t \Phi_{t,t_i} \, dt$$

To ease the notation we define the matrix function

$$S_{k_t} := D_t + \int_{kh}^{t} C_t \Phi_{t,\tau} B_\tau \, d\tau$$

Furthermore $M_k := M_{kh,\infty}$, $\boldsymbol{M} := \operatorname{diag}\{M_k\}$ and $\boldsymbol{N} := \operatorname{diag}\{N_k\}$ where

$$N_k := \int_{kh}^{(k+1)h} S_{k_t}^T C_t \Phi_{t,kh} \, dt$$

With these definitions, we are ready to state the following:

Theorem 3.13 *The norm of $G\boldsymbol{H}_h : \ell_2^m \to \mathcal{L}_2^p$ equals $\|\boldsymbol{Q}\|^{1/2}$, where $\boldsymbol{Q}$ is the discrete-time operator*

$$\boldsymbol{Q} = \boldsymbol{R} + \boldsymbol{R}^* + \Lambda$$

Here $\boldsymbol{R}$ is a causal operator with state-space representation

$$\Sigma_R := \left[\begin{array}{c|c} \boldsymbol{A} & \boldsymbol{B} \\ \hline \boldsymbol{C} & \boldsymbol{0} \end{array}\right] \tag{3.11}$$

The state-space operators are given by

$$\boldsymbol{A} := \operatorname{diag}\{\Phi((k+1)h, kh)\}$$
$$\boldsymbol{B} := \operatorname{diag}\{B_k\}; \qquad B_k := \int_{kh}^{(k+1)h} \Phi((k+1)h, \tau) B(\tau) \, d\tau$$
$$\boldsymbol{C} := \boldsymbol{N} + \boldsymbol{B}^* \boldsymbol{Z} \boldsymbol{M} \boldsymbol{Z}^* \boldsymbol{A}$$

and the memoryless operator

$$\Lambda := \operatorname{diag}\left\{ B_k^T M_{k+1} B_k + \int_{kh}^{(k+1)h} S_{k_t}^T S_{k_t}\, dt \right\} \tag{3.12}$$

Proof: First notice that, using the boundedness assumptions placed on the state-space matrices of (3.10), it is easy to show that Λ is bounded. Also notice that $\boldsymbol{R}$ is stable since $\boldsymbol{A}$ is UES (Remark 3.6). Recall that $\boldsymbol{T} = \boldsymbol{G}\boldsymbol{H}_h$ and $\boldsymbol{Q} := \boldsymbol{T}^*\boldsymbol{T}$. Since $\|\boldsymbol{T}\| = \|\boldsymbol{T}^*\boldsymbol{T}\|^{1/2}$, all that remains to show is that $\boldsymbol{Q}$ is given by the expression above. Let $r = \boldsymbol{T}\boldsymbol{w}$; then

$$\langle r, r\rangle_{\mathcal{L}_2^p} = \langle \boldsymbol{w}, \boldsymbol{Q}\boldsymbol{w}\rangle_{\ell_2^m} \tag{3.13}$$

Also

$$\begin{aligned} r_t &= \int_{-\infty}^{t} g_{t,\tau} w_\tau \, d\tau \\ &= \sum_{k=-\infty}^{\infty} H_{k_t} w_k \end{aligned} \tag{3.14}$$

where the matrix function

$$H_{k_t} := \begin{cases} 0 & \text{for } t < kh \\ S_{k_t} & \text{for } kh \le t < (k+1)h \\ C_t \Phi_{t,(k+1)h} B_k & \text{for } (k+1)h \le t \end{cases} \tag{3.15}$$

From (3.13), (3.14) and (3.15) we have that

$$Q_{k,l} = \int_{-\infty}^{\infty} H_{k_t}^T H_{l_t}\, dt$$

Now, it is easy to check that, for k equal to l, the main diagonal of $\boldsymbol{Q}$ is given by Λ in (3.12). We now show that the lower triangular part of $\boldsymbol{Q}$ is $\boldsymbol{R}$. Since $\boldsymbol{Q}$ is a self-adjoint operator, the upper triangular part is then given by $\boldsymbol{R}^*$.

Let $k \geq l+1$; then we can evaluate

$$\begin{aligned}
Q_{k,l} &= \int_{kh}^{(k+1)h} S_{k_t}^T C_t \Phi_{t,(l+1)h}\, dt\, B_l \\
&\quad + B_k^T \int_{(k+1)h}^{\infty} \Phi_{t,(k+1)h}^T C_t^T C_t \Phi_{t,(l+1)h}\, dt\, B_l \\
&= \int_{kh}^{(k+1)h} S_{k_t} C_t \Phi_{t,kh}\, dt\, \Phi_{kh,(l+1)h} B_l \\
&\quad + B_k^T \int_{(k+1)h}^{\infty} \Phi_{t,(k+1)h}^T C_t^T C_t \Phi_{t,(k+1)h}\, dt A_k \Phi_{kh,(l+1)h} B_l \\
&= \left[N_k + B_k^T M_{k+1} A_k \right] \Phi_{kh,(l+1)h} B_l
\end{aligned}$$

Since

$$\begin{aligned}
\Phi_{kh,(l+1)h} &= \Phi_{kh,(k-1)h} \Phi_{(k-1)h,(k-2)h} \cdots \Phi_{(l+2)h,(l+1)h} \\
&= A_{k-1} A_{k-2} \cdots A_{l+1}
\end{aligned}$$

and

$$\boldsymbol{N} + \boldsymbol{B}^* \boldsymbol{Z} \boldsymbol{M} \boldsymbol{Z}^* \boldsymbol{A} = \operatorname{diag}\left\{ N_k + B_k^T M_{k+1} A_k \right\}$$

we have that

$$Q_{k,l} = C_k A_{k-1} \cdots A_{l+1} B_l$$

which is the impulse response of the discrete-time system given by (3.11). ∎

As in Section 3.2.1, we first compute a factorization $\boldsymbol{Q} = \boldsymbol{P}^* \boldsymbol{P}$, and then apply Theorem 3.1 to the factor $\boldsymbol{P}$. Note that this is the dual version. If $\boldsymbol{Q} > \boldsymbol{0}$ (see Remark 3.10 for the case where this is not satisfied), a factorization $\boldsymbol{Q} = \boldsymbol{P}^* \boldsymbol{P}$, where $\boldsymbol{P}$ is causal, is given in [78] by the state-space realization

$$\left[\begin{array}{c|c} \boldsymbol{A} & \boldsymbol{B} \\ \hline (\Lambda - \boldsymbol{B}^* \boldsymbol{Z} \boldsymbol{S} \boldsymbol{Z}^* \boldsymbol{B})^{-1/2} (\boldsymbol{C} - \boldsymbol{B}^* \boldsymbol{Z} \boldsymbol{S} \boldsymbol{Z}^* \boldsymbol{A}) & (\Lambda - \boldsymbol{B}^* \boldsymbol{Z} \boldsymbol{S} \boldsymbol{Z}^* \boldsymbol{B})^{1/2} \end{array} \right]$$

where $\boldsymbol{0} \leq \boldsymbol{S} \in \mathcal{M}$ is the solution of

$$\begin{aligned}
\boldsymbol{S} &= \boldsymbol{A}^* \boldsymbol{Z} \boldsymbol{S} \boldsymbol{Z}^* \boldsymbol{A} \\
&\quad + (\boldsymbol{A}^* \boldsymbol{Z} \boldsymbol{S} \boldsymbol{Z}^* \boldsymbol{B} - \boldsymbol{C}^*)(\Lambda - \boldsymbol{B}^* \boldsymbol{Z} \boldsymbol{S} \boldsymbol{Z}^* \boldsymbol{B})^{-1} (\boldsymbol{B}^* \boldsymbol{Z} \boldsymbol{S} \boldsymbol{Z}^* \boldsymbol{A} - \boldsymbol{C})
\end{aligned} \tag{3.16}$$

such that $\Lambda - \boldsymbol{B}^* \boldsymbol{Z} \boldsymbol{S} \boldsymbol{Z}^* \boldsymbol{B} > \boldsymbol{0}$. Since $\boldsymbol{P}$ is causal, we can find conditions for the norm bound on $\boldsymbol{Q}$ in terms of this factor. The proof is similar to the proof of Theorem 3.9, and will therefore be omitted.

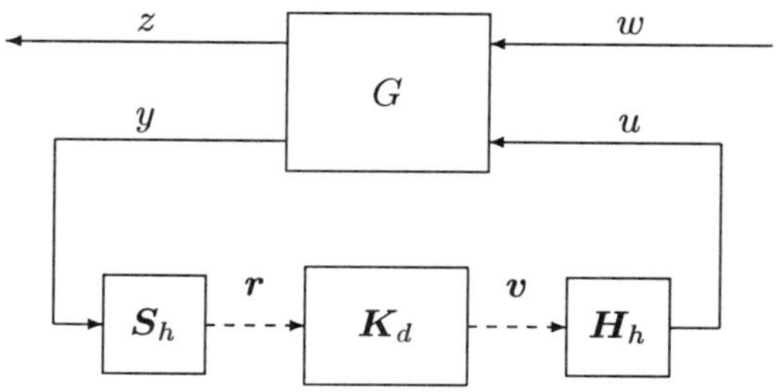

Figure 3.3: Hybrid feedback system

Theorem 3.14 *Suppose that $\boldsymbol{Q} > \mathbf{0}$. Then the following statements are equivalent:*

1. $\boldsymbol{I} - \boldsymbol{Q} > 0$

2. *There exists $\mathbf{0} \leq \boldsymbol{X} \in \mathcal{M}$ satisfying the operator Riccati equation*

$$\boldsymbol{X} = \boldsymbol{A}^*\boldsymbol{Z}\boldsymbol{X}\boldsymbol{Z}^*\boldsymbol{A} + (\boldsymbol{A}^*\boldsymbol{Z}\boldsymbol{X}\boldsymbol{Z}^*\boldsymbol{B} + \boldsymbol{C}^*)\,\boldsymbol{V}^{-1}\,(\boldsymbol{B}^*\boldsymbol{Z}\boldsymbol{X}\boldsymbol{Z}^*\boldsymbol{A} + \boldsymbol{C}) \tag{3.17}$$

with $\boldsymbol{V} > \mathbf{0}$ such that $\boldsymbol{A}_F$ is UES. Here

$$\begin{aligned} \boldsymbol{V} &:= \boldsymbol{I} - \Lambda - \boldsymbol{B}^*\boldsymbol{Z}\boldsymbol{X}\boldsymbol{Z}^*\boldsymbol{B} \\ \boldsymbol{A}_F &:= \boldsymbol{A} + \boldsymbol{B}\boldsymbol{V}^{-1}(\boldsymbol{B}^*\boldsymbol{Z}\boldsymbol{X}\boldsymbol{Z}^*\boldsymbol{A} + \boldsymbol{C}) \end{aligned}$$

3. *There exists a spectral factorization $\boldsymbol{I} - \boldsymbol{Q} = \boldsymbol{M}^*\boldsymbol{M}$ for some operator $\boldsymbol{M} \in \mathcal{C} \cap \mathcal{C}^{-1}$.* □

3.2.3 Hybrid feedback systems

In this section we show how the results of Section 3.1 can be used to characterize the norm of a hybrid feedback system. Consider the feedback system of Figure 3.3. The continuous-time plant G is given by a time-varying state-space representation:

$$\Sigma_G := \left\{ \begin{array}{rcll} \dot{x}_t &=& A_t x_t + B_{1_t} w_t & +B_{2_t} u_t \\ z_t &=& C_{1_t} x_t & +D_{12_t} u_t \\ y_t &=& C_{2_t} x_t & \end{array} \right.$$

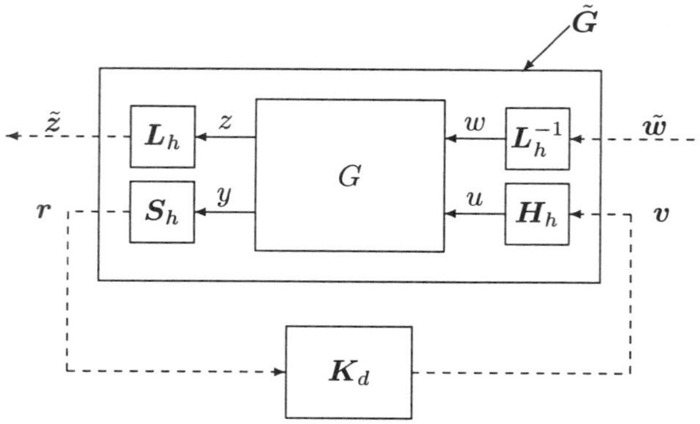

Figure 3.4: "Lifted" feedback system.

and is controlled by a continuous-time controller K which is obtained from a discrete-time controller $\boldsymbol{K}_d$:

$$\Sigma_{K_d} := \left[\begin{array}{c|c} \boldsymbol{A}_K & \boldsymbol{B}_K \\ \hline \boldsymbol{C}_K & \boldsymbol{D}_K \end{array}\right]$$

via sample-and-hold circuits. We are interested in a characterization of the norm of the feedback system:

$$\begin{aligned} T &:= G_{11} + G_{12}K\big(I - G_{22}K\big)^{-1}G_{21} \\ &= G_{11} + G_{12}\boldsymbol{H}_h\boldsymbol{K}_d\big(\boldsymbol{I} - \boldsymbol{S}_h G_{22}\boldsymbol{H}_h\boldsymbol{K}_d\big)^{-1}\boldsymbol{S}_h G_{21} \end{aligned}$$

The operator T maps continuous-time signals to continuous-time signals, although it has components which are purely discrete: $\boldsymbol{S}_h G_{22}\boldsymbol{H}_h$ and $\boldsymbol{K}_d$; purely continuous: G_{11}; as well as hybrid: $G_{12}\boldsymbol{H}_h$ and $\boldsymbol{S}_h G_{21}$. We can use the lifting techniques introduced in [11, 12] to compute the norm of the closed-loop system T. We introduce the lifting operator $\boldsymbol{L}_h$ which maps a continuous-time signal to a discrete signal according to

$$\tilde{\boldsymbol{L}} = \boldsymbol{L}_h y \iff \tilde{\boldsymbol{L}} = \{\tilde{y}_k\}_{k=-\infty}^{\infty} \text{ with } \tilde{y}_k(t) = y(t+kh), \quad t \in [0,h)$$

The elements of $\tilde{\boldsymbol{L}}$ are in the compressed space $\mathcal{K} := \mathcal{L}_2^n[0,h)$. Signals $y \in \mathcal{L}_2^n$ are lifted to signals $\tilde{y} \in \ell_2^n(\mathcal{K})$. Using the lifted signals $\tilde{\boldsymbol{z}} := \boldsymbol{L}_h z$ and $\tilde{\boldsymbol{w}} := \boldsymbol{L}_h w$ we obtain the closed-loop system of Figure 3.4

$$\tilde{\boldsymbol{T}} := \tilde{\boldsymbol{G}}_{11} + \tilde{\boldsymbol{G}}_{12}\boldsymbol{K}_d(\boldsymbol{I} - \tilde{\boldsymbol{G}}_{22}\boldsymbol{K}_d)^{-1}\tilde{\boldsymbol{G}}_{21}$$

where

$$\tilde{\boldsymbol{G}} := \begin{bmatrix} \tilde{\boldsymbol{G}}_{11} & \tilde{\boldsymbol{G}}_{12} \\ \tilde{\boldsymbol{G}}_{21} & \tilde{\boldsymbol{G}}_{22} \end{bmatrix} := \begin{bmatrix} \boldsymbol{L}_h G_{11} \boldsymbol{L}_h^{-1} & \boldsymbol{L}_h G_{12} \boldsymbol{H}_h \\ \boldsymbol{S}_h G_{21} \boldsymbol{L}_h^{-1} & \boldsymbol{S}_h G_{22} \boldsymbol{H}_h \end{bmatrix}$$

A state space representation for the system $\tilde{\boldsymbol{G}}$ is given by

$$\Sigma_{\tilde{G}} := \left[\begin{array}{c|cc} \tilde{\boldsymbol{A}} & \tilde{\boldsymbol{B}}_1 & \tilde{\boldsymbol{B}}_2 \\ \hline \tilde{\boldsymbol{C}}_1 & \tilde{\boldsymbol{D}}_{11} & \tilde{\boldsymbol{D}}_{12} \\ \tilde{\boldsymbol{C}}_2 & \boldsymbol{0} & \boldsymbol{0} \end{array}\right]$$

which maps $\ell_2^{m_1}(\mathcal{K}) \oplus \ell_2^{m_2}$ to $\ell_2^{p_1}(\mathcal{K}) \oplus \ell_2^{p_2}$. Here

$$\begin{aligned} \tilde{\boldsymbol{A}} &= \operatorname{diag}\{\Phi_{(k+1)h,kh}\} \\ \tilde{\boldsymbol{B}}_2 &= \operatorname{diag}\left\{\int_{kh}^{(k+1)h} \Phi_{(k+1)h,\tau} B_{2_\tau}\, d\tau\right\} \\ \tilde{\boldsymbol{C}}_2 &= \operatorname{diag}\{C_{2_{kh}}\} \end{aligned}$$

are standard discrete-time elements; $\tilde{\boldsymbol{B}}_1 = \operatorname{diag}\left\{\tilde{B}_{1_k}\right\}$ is a diagonal operator with the kth element mapping $\mathcal{K}^{m_1}$ to $\mathbb{R}^n$ according to

$$\tilde{B}_{1_k}\tilde{w}_k = \int_{kh}^{(k+1)h} \Phi_{(k+1)h,\tau} B_{1_\tau} w_\tau\, d\tau$$

Similarly, $\tilde{\boldsymbol{C}}_1$ and $\tilde{\boldsymbol{D}}_{12}$ are diagonal operators with elements

$$\begin{aligned} \tilde{C}_{1_k} &: \mathbb{R}^n \to \mathcal{K}^{p_1}, \quad \tilde{C}_{1_{k_t}} = C_{1_t}\Phi_{t,kh} \\ \tilde{D}_{12_k} &: \mathbb{R}^{m_2} \to \mathcal{K}^{p_1}, \quad \tilde{D}_{12_{k_t}} = \left(D_{12_t} + C_{1_t}\int_{kh}^{t} \Phi_{t,\tau} B_{2_\tau}\, d\tau\right) \end{aligned}$$

for $t \in [kh, (k+1)h)$. Finally, $\tilde{\boldsymbol{D}}_{11} = \operatorname{diag}\left\{\tilde{D}_{11_k}\right\}$ is the operator

$$\left(\tilde{D}_{11_k}\tilde{w}_k\right)_t = C_{1_t}\int_{kh}^{t} \Phi_{t,\tau} B_{1_\tau} w_\tau\, d\tau, \qquad t \in [kh, (k+1)h)$$

Using these operators, we obtain a state-space representation for the closed-loop system

$$\Sigma_{\tilde{T}} = \left[\begin{array}{c|c} \boldsymbol{A}_{cl} & \boldsymbol{B}_{cl} \\ \hline \boldsymbol{C}_{cl} & \boldsymbol{D}_{cl} \end{array}\right] := \left[\begin{array}{cc|c} \tilde{\boldsymbol{A}} + \tilde{\boldsymbol{B}}_2 \boldsymbol{D}_K \tilde{\boldsymbol{C}}_2 & \tilde{\boldsymbol{B}}_2 \boldsymbol{C}_K & \tilde{\boldsymbol{B}}_1 \\ \boldsymbol{B}_K \tilde{\boldsymbol{C}}_2 & \boldsymbol{A}_K & \boldsymbol{0} \\ \hline \tilde{\boldsymbol{C}}_1 + \tilde{\boldsymbol{D}}_{12} \boldsymbol{D}_K \tilde{\boldsymbol{C}}_2 & \tilde{\boldsymbol{D}}_{12} \boldsymbol{C}_K & \boldsymbol{0} \end{array}\right]$$

From [12] we know that the induced norm of $T : \mathcal{L}_2^{m_1} \to \mathcal{L}_2^{m_1}$ is the same as that of $\tilde{\boldsymbol{T}} : \ell_2^{m_1}(\mathcal{K}) \to \ell_2^{m_1}(\mathcal{K})$. The norm of this operator can be computed as in Section 3.1 giving the following result:

Theorem 3.15 $\|\tilde{\boldsymbol{T}}\| < 1$ *if and only if there exists* $\boldsymbol{0} \leq \boldsymbol{X} \in \mathcal{M}$ *satisfying the operator Riccati equation*

$$\boldsymbol{X} = \boldsymbol{A}_{cl}^* \boldsymbol{Z} \boldsymbol{X} \boldsymbol{Z}^* \boldsymbol{A}_{cl} + \boldsymbol{C}_{cl}^* \boldsymbol{C}_{cl} + \boldsymbol{A}_{cl}^* \boldsymbol{Z} \boldsymbol{X} \boldsymbol{Z}^* \boldsymbol{B}_{cl} \boldsymbol{V}_{cl}^{-1} \boldsymbol{B}_{cl}^* \boldsymbol{Z} \boldsymbol{X} \boldsymbol{Z}^* \boldsymbol{A}_{cl} \tag{3.18}$$

with $\boldsymbol{V}_{cl} := \boldsymbol{I} - \boldsymbol{B}_{cl}^* \boldsymbol{Z} \boldsymbol{X} \boldsymbol{Z}^* \boldsymbol{B}_{cl} > \boldsymbol{0}$ *such that* $\boldsymbol{A}_{cl} + \boldsymbol{B}_{cl} \boldsymbol{V}_{cl}^{-1} \boldsymbol{B}_{cl}^* \boldsymbol{Z} \boldsymbol{X} \boldsymbol{Z}^* \boldsymbol{A}_{cl}$ *is UES.*

Proof: This follows easily by noting that the lifting operation is norm-preserving, and Theorem 3.1. ∎

Remark 3.16 Theorem 3.15 can be used to obtain the induced-norm results of Theorem 3.9 by setting $\boldsymbol{K}_d = \boldsymbol{I}$ and

$$\begin{bmatrix} G_{11} & G_{12} \\ G_{21} & G_{22} \end{bmatrix} = \begin{bmatrix} 0 & I \\ G & 0 \end{bmatrix}$$

Nevertheless, the Riccati equation associated with Theorem 3.9 has the advantage over the Riccati equation (3.18) in that all of the state-space operators of (3.8) are block-diagonal operators with matrix elements. Some of the elements of (3.18) (e.g. $\tilde{\boldsymbol{B}}_1$ and $\tilde{\boldsymbol{C}}_1$) consist of block-diagonal operators where the individual blocks are not finite-dimensional matrices.

In the next section we will look at some computational aspects of the different Riccati operator equations considered in this chapter.

3.3 Computational issues

In order to apply some of the results presented in this paper one is faced with the difficulty in trying to compute solutions to the operator Riccati equations. These equations can be separated into two main groups: non-causal equations (3.1), (3.16), (3.17), (3.18); and causal equations (3.2), (3.7), (3.8), (3.9). This designation is based on the order in which the two shifts, $\boldsymbol{Z}$ and $\boldsymbol{Z}^*$, appear in the right-hand side of the equation.

For causal operator Riccati equations, the appearance of $\boldsymbol{Z}^* \boldsymbol{Y} \boldsymbol{Z}$ has the effect of shifting the elements of the diagonal operator $\boldsymbol{Y}$ one element "down." Thus, for example, the operator Riccati equation (3.2) could be written as a Riccati difference equation evolving forwards in time

$$Y_k = A_k Y_{k-1} A_k^T + B_k B_k^T + \bar{R}_k^T (I - D_k D_k^T - C_k Y_{k-1} C_k^T)^{-1} \bar{R}_k$$

where $\bar{R}_k = C_k Y_{k-1} A_k^T + D_k B_k^T$.

For non-causal Riccati equations of the type (3.1), the action of $\boldsymbol{Z X Z}^*$ is equivalent to the shifting of the elements of the diagonal operator $\boldsymbol{X}$ one step "up." Consequently, the resulting Riccati difference equation is now evolving backwards in time

$$X_k = A_k^T X_{k+1} A_k + C_k^T C_k + R_k^T (I - D_k^T D_k - B_k^T X_{k+1} B_k)^{-1} R_k$$

where $R_k = B_k^T X_{k+1} A_k + D_k^T C_k$. There are practical difficulties for computing these solutions for systems on an infinite-time horizon. Even for systems on a finite time horizon where the time-dependence is not known *a priori*, this will require approximate solutions.

In case we work with single-infinite sequences, equation (3.2) has an advantage over (3.1). Namely X_k has to be computed for $k \geq 0$ with no initial condition, while Y_k can be computed straightforward for $k \geq 0$ using the recursive expression with initial condition $Y_{-1} = 0$.

For LTI systems evolving on a doubly-infinite time axis, the operators $\boldsymbol{X}$ and $\boldsymbol{Y}$ both commute with the two shifts. As such, $\boldsymbol{Z X Z}^* = \boldsymbol{X}$ and $\boldsymbol{Z}^* \boldsymbol{Y Z} = \boldsymbol{Y}$ and so the operator Riccati equations are now just the usual matrix algebraic Riccati equations for discrete-time LTI systems along the diagonal. This observation leads one to try to compute an approximation of the operators $\boldsymbol{X}$ by making the substitution $\boldsymbol{Z X Z}^* \approx \boldsymbol{X}$ in the LTV case. Intuitively this should give approximate solutions when $\|\boldsymbol{Z X Z}^* - \boldsymbol{X}\|$ is small; that is, when the system is slowly time-varying. This approach has been used repeatedly in the literature for the design of linear quadratic regulators in the LTV case [28, 38, 60]. Related results on this slowly-time varying approach have been considered by Zames and Wang [79, 80, 86]. For general LTV systems, however, computing the solution to the operator Riccati equations remains an open problem.

4 Discrete-Time Entropy

In Chapter 1 it was shown that controllers which minimize the entropy of the closed-loop system provide a simple means of trading off the robustness and performance that is usually associated with the $\mathcal{H}_\infty$ and $\mathcal{H}_2$ cost functions for linear time-invariant systems.

This property makes it desirable to extend the notion of entropy to a broader class of systems than those which admit a transfer function representation. In this chapter, it will be shown that for discrete-time time-varying systems, a natural extension of the entropy exists. The definition of the entropy for this broader class of systems will be based on the spectral factorization problem that was considered in Chapter 3.

The rest of the chapter will be devoted to deriving properties of this entropy. In particular, it will be shown that many of the properties of the entropy integral that is used for time-invariant systems carry over to the new setting. The one exception concerns the relationship between the entropy of a causal system and that of its (anti-causal) adjoint. For linear time-invariant systems, the two entropies are equal to each other, and this fact is used repeatedly in solving the minimum entropy control problem. For time-varying systems it will be shown that in general, the two are not equal. For that reason we will also introduce the notion of entropy for an anti-causal system, and discuss the relationship between the two entropies.

4.1 Entropy of a discrete-time time-varying system

In Chapter 2, a representation for discrete-time time-varying systems based on infinite-dimensional operators was introduced. This representation was used in Theorem 3.1 to characterize the induced norm of such an operator in terms of a spectral factorization. In the time-invariant case, the entropy can also be expressed in terms of this factorization. Recall the definition of the entropy of an LTI system

$$I_d(G,\gamma) = -\frac{\gamma^2}{2\pi}\int_{-\pi}^{\pi} \ln\left|\det\left(I-\gamma^{-2}G(e^{-i\omega})^T G(e^{i\omega})\right)\right| d\omega \tag{4.1}$$

Since $I-\gamma^{-2}G(z)^{\sim}G(z)$ is a positive function, where $G(z)^{\sim} = G(\bar{z}^{-1})^H$ is the para-Hermitian conjugate, it is possible to compute a spectral factorization

$$I-\gamma^{-2}G(z)^{\sim}G(z) = M(z)^{\sim}M(z)$$

where $M(z) \in \mathcal{H}_\infty$ and $M(z)^{-1} \in \mathcal{H}_\infty$. Hence, with $G(e^{i\omega})^{\sim} = G(e^{-i\omega})^T$, we can write (4.1) as

$$\begin{aligned} I_d(G,\gamma) &= -\frac{\gamma^2}{2\pi}\int_{-\pi}^{\pi} \ln\det\left(I-\gamma^{-2}G(e^{-i\omega})^T G(e^{i\omega})\right) d\omega \\ &= -\frac{\gamma^2}{2\pi}\int_{-\pi}^{\pi} \ln\det\left(M(e^{-i\omega})^T M(e^{i\omega})\right) d\omega \\ &= -\frac{\gamma^2}{\pi}\int_{-\pi}^{\pi} \ln\left|\det\left(M(e^{i\omega})\right)\right| d\omega \end{aligned}$$

Since $\ln|\det(M(z))|$ is outer, we may use the Poisson integral formula [69, Theorem 17.16] to simplify this as

$$\begin{aligned} I_d(G,\gamma) &= -2\gamma^2\ln|\det(M(\infty))| \\ &= -\gamma^2\ln\det\left(M(\infty)^T M(\infty)\right) \end{aligned} \tag{4.2}$$

It follows that the entropy of the transfer function $G(z)$ can be expressed in terms of the memoryless part of the spectral factor $M(z)$.

Based on this interpretation of the entropy, we now consider time-varying systems. As in Section 2.1, we can represent a linear discrete-time time-varying system as an infinite-dimensional operator $\boldsymbol{G}$. Suppose that $\boldsymbol{G}$ is causal with induced norm $\|\boldsymbol{G}\| < \gamma$. It follows that the self-adjoint operator $\boldsymbol{I}-\gamma^{-2}\boldsymbol{G}^*\boldsymbol{G}$ is positive. By Lemma 2.4, it has a spectral factorization.

Specifically, $\boldsymbol{M} \in \mathcal{C} \cap \mathcal{C}^{-1}$ satisfies

$$\boldsymbol{I} - \gamma^{-2}\boldsymbol{G}^*\boldsymbol{G} = \boldsymbol{M}^*\boldsymbol{M}$$

Using the memoryless part of the spectral factor $\boldsymbol{M}$, we can define an entropy operator mimicking the right-hand side of (4.2).

Definition 4.1 *Suppose that $\boldsymbol{G} \in \mathcal{C}$ and $\|\boldsymbol{G}\| < \gamma$. Let $\boldsymbol{M} \in \mathcal{C} \cap \mathcal{C}^{-1}$ be a spectral factor of the positive operator $\boldsymbol{I} - \gamma^{-2}\boldsymbol{G}^*\boldsymbol{G}$. We define the entropy by*

$$\boldsymbol{E}(\boldsymbol{G}, \gamma) := -\gamma^2 \operatorname{diag}\left\{\ln\det\left(M_{k,k}^T M_{k,k}\right)\right\} \in \mathcal{M} \tag{4.3}$$

Without loss of generality, we will typically take $\gamma = 1$, and use the notation $\boldsymbol{E}(\boldsymbol{G}) := \boldsymbol{E}(\boldsymbol{G}, 1)$.

Remark 4.2 The entropy is defined as a memoryless operator. We will use the terminology "entropy at time k" to denote the kth diagonal element of this operator.

Although spectral factors are not unique, it can be shown that the entropy does not depend on the particular choice of spectral factor. Recall from Lemma 2.4 that any other spectral factor can be written as $\boldsymbol{UM}$, for some $\boldsymbol{U} \in \mathcal{M} \cap \mathcal{M}^{-1}$ such that $\boldsymbol{U}^*\boldsymbol{U} = \boldsymbol{I}$. Hence $U_{k,k}^T U_{k,k} = I$, and it is obvious that the entropy is not affected by U. Furthermore, since $\boldsymbol{M}$ is a causal operator with a causal inverse, Lemma 2.3 shows that $\operatorname{diag}\{M_{k,k}^T M_{k,k}\} \geq \epsilon \boldsymbol{I}$ for some $\epsilon > 0$, ensuring that the entropy operator is well-defined.

Remark 4.3 In contrast to (4.1), where entropy is defined as a real number, the entropy considered here is a memoryless operator where the diagonal elements are positive real numbers. This will require that the minimization of entropy be carried out with respect to the partial ordering of positive operators. This poses no difficulty. Moreover, it is reminiscent of the stochastic optimization problem considered in [36].

Remark 4.4 The entropy (4.3) differs from that used in [51] where entropy was defined as

$$\operatorname{diag}\{M_{k,k}^T M_{k,k}\} \tag{4.4}$$

Consider an operator $\boldsymbol{N}$ satisfying

$$\boldsymbol{I} \geq \operatorname{diag}\{M_{k,k}^T M_{k,k}\} \geq \operatorname{diag}\{N_{k,k}^T N_{k,k}\} > \boldsymbol{0}$$

It follows that, for all k

$$0 \le -\gamma^2 \ln \det(M_{k,k}^T M_{k,k}) \le -\gamma^2 \ln \det(N_{k,k}^T N_{k,k}) < \infty$$

Hence, maximizing entropy operators of the form $\mathrm{diag}\{M_{k,k}^T M_{k,k}\}$ implies minimizing the entropy (4.3). The modification introduced here is needed because there are instances where the entropy cannot be optimized with respect to the partial ordering of (4.4) but may be optimized with respect to the partial ordering of (4.3). For the full information problem considered in Section 6.3, either definition can be used; it is only when we consider more general problems that we require the extra ordering introduced by (4.3). As an example consider the set of operators

$$\boldsymbol{G} = \mathrm{diag}\left\{\left[\begin{array}{cc} \frac{1}{2} & a \end{array}\right]\right\}$$

for $a^2 < \frac{3}{4}$. Since $\boldsymbol{G}$ is memoryless, it is immediate that, with $\gamma = 1$

$$M_{k,k}^T M_{k,k} = I - G_{k,k}^T G_{k,k} = \left[\begin{array}{cc} \frac{3}{4} & -\frac{a}{2} \\ -\frac{a}{2} & 1 - a^2 \end{array}\right]$$

This cannot be maximized with respect to the partial ordering of (4.4). However, since the determinant of this matrix equals $\frac{3}{4} - a^2$, minimization with respect to the partial ordering of (4.3) is possible, with the optimal value $a = 0$.

In the next section we will discuss some useful properties of the entropy operator.

4.2 Properties

In this section we present some properties of the entropy operator. We will show that, in case the system is time-invariant, the usual entropy can be recovered from our notion of entropy. Also, for systems which admit a state-space realization, we will express the entropy in terms of the state-space operators.

We begin by showing that the entropy has some norm-like features. In particular, it is non-negative, and equals the zero operator if and only if $\boldsymbol{G}$ is the zero operator. This property is analogous to that found in the time-invariant case, and will be used repeatedly in solving the minimum entropy control problem.

Lemma 4.5 *Suppose that* $\boldsymbol{G} \in \mathcal{C}$ *with* $\|\boldsymbol{G}\| < \gamma$. *Then* $\boldsymbol{E}(\boldsymbol{G}, \gamma) \geq \boldsymbol{0}$ *with equality if and only if* $\boldsymbol{G} = \boldsymbol{0}$.

Proof: From Lemma 2.3 we know that $M_{k,k}^T M_{k,k}$ is positive. Moreover, since

$$M_{k,k}^T M_{k,k} = I - \gamma^{-2}(\boldsymbol{G}^*\boldsymbol{G})_{k,k} - \sum_{i=k+1}^{\infty} M_{i,k}^T M_{i,k} \leq I \tag{4.5}$$

it is immediate that $\boldsymbol{E}(\boldsymbol{G}, \gamma) \geq \boldsymbol{0}$.

That $\boldsymbol{E}(\boldsymbol{0}, \gamma) = \boldsymbol{0}$ is straightforward. Now suppose that $\boldsymbol{E}(\boldsymbol{G}, \gamma) = \boldsymbol{0}$. Combining (4.5) with Lemma 2.3 shows that $0 < M_{k,k}^T M_{k,k} \leq I$ for all k. Since the entropy equals zero, we know that $M_{k,k}^T M_{k,k} = I$ for all k. Take any $l \geq k$. Then

$$\begin{aligned} I - \gamma^{-2} G_{l,k}^T G_{l,k} &\geq I - \gamma^{-2} \sum_{i=k}^{\infty} G_{i,k}^T G_{i,k} \\ &= \sum_{i=k}^{\infty} M_{i,k}^T M_{i,k} \\ &\geq M_{k,k}^T M_{k,k} \\ &= I \end{aligned}$$

It follows that $G_{l,k} = 0$ for all $l \geq k$, which is equivalent to $\boldsymbol{G}$ being the zero operator. ■

Remark 4.6 Since $0 < M_{k,k}^T M_{k,k} \leq I$, all the eigenvalues of $M_{k,k}^T M_{k,k}$ are in the interval $(0, 1]$. Using this, the result of Lemma 4.5 can be restated as

$$\operatorname{diag}\left\{\det\left(M_{k,k}^T M_{k,k}\right)\right\} \leq \boldsymbol{I}$$

with equality if and only if $\boldsymbol{M}^*\boldsymbol{M} = \boldsymbol{I}$. Moreover, minimizing the entropy at time k is equivalent to maximizing the product of the eigenvalues of $M_{k,k}^T M_{k,k}$.

Another important property is that the entropy is decreasing with respect to the parameter γ. Although this result is similar to the time-invariant case, the proof is much harder. For time-invariant systems, the entropy expression (4.1) in terms of the transfer function can easily be evaluated with a power series expansion, from which the result follows immediately.

Lemma 4.7 *For $\gamma \geq \bar{\gamma} > \|\boldsymbol{G}\|$, we have that $\boldsymbol{E}(\boldsymbol{G},\gamma) \leq \boldsymbol{E}(\boldsymbol{G},\bar{\gamma})$.*

Proof: For fixed k, denote the eigenvalues of the $m \times m$ matrix $M_{k,k}^T M_{k,k}$ by $\lambda_1, \cdots, \lambda_m$. The entropy at time k can be evaluated as

$$\begin{aligned}
\boldsymbol{E}(\boldsymbol{G},\gamma)_k &= -\gamma^2 \ln \det(M_{k,k}^T M_{k,k}) \\
&= -\gamma^2 \ln \prod_{i=1}^{m} \lambda_i \\
&= -\gamma^2 \sum_{i=1}^{m} \ln \lambda_i \\
&= -\gamma^2 \sum_{i=1}^{m} \ln(1-(1-\lambda_i))
\end{aligned} \tag{4.6}$$

Since all the eigenvalues are in the interval $(0, 1]$, we may carry out a power series expansion of the natural logarithm, i.e.

$$\ln(1-x) = \sum_{i=1}^{\infty} \frac{x^j}{j} \qquad \text{if } |x| < 1$$

Applying this expansion to (4.6) yields

$$\begin{aligned}
\boldsymbol{E}(\boldsymbol{G},\gamma)_k &= \gamma^2 \sum_{i=1}^{m} \sum_{j=1}^{\infty} \frac{(1-\lambda_i)^j}{j} \\
&= \gamma^2 \sum_{j=1}^{\infty} \frac{1}{j} \sum_{i=1}^{m} (1-\lambda_i)^j \\
&= \gamma^2 \sum_{j=1}^{\infty} \frac{1}{j} \operatorname{tr}\left[(I - M_{k,k}^T M_{k,k})^j\right] \\
&= \sum_{j=1}^{\infty} \frac{\gamma^{-2(j-1)}}{j} \operatorname{tr}\left[(\gamma^2 I - \gamma^2 M_{k,k}^T M_{k,k})^j\right]
\end{aligned} \tag{4.7}$$

Obviously, for every $j \geq 1$, the term $\gamma^{-2(j-1)}$ is a decreasing function with respect to γ. Consider the second term in (4.7). Suppose that $\gamma \geq \bar{\gamma} > \|\boldsymbol{G}\|$, and denote the spectral factor corresponding to $\bar{\gamma}$ by $\boldsymbol{L}$. That is

$$\gamma^2 I - \gamma^2 \boldsymbol{M}^* \boldsymbol{M} = \boldsymbol{G}^* \boldsymbol{G} = \bar{\gamma}^2 I - \bar{\gamma}^2 \boldsymbol{L}^* \boldsymbol{L} \tag{4.8}$$

Recalling from Lemma 2.3 that $M_{i,i}$ is invertible, we may define a signal

$\boldsymbol{w} = \{w_i\}$ by

$$w_i = \begin{cases} 0 & \text{if } i < k \\ \bar{w} & \text{if } i = k \\ -M_{i,i}^{-1} \sum_{j=k}^{i-1} M_{i,j} w_j & \text{if } i > k \end{cases}$$

where $\bar{w}$ is any unit-length vector. Notice that $(\boldsymbol{M}\boldsymbol{w})_i = 0$ for all $i \neq k$. Pre- and post-multiplying (4.8) by this signal gives

$$\begin{aligned} \gamma^2 \|\boldsymbol{w}\|_2^2 - \gamma^2 |M_{k,k}\bar{w}|^2 &= \bar{\gamma}^2 \|\boldsymbol{w}\|_2^2 - \bar{\gamma}^2 |L_{k,k}\bar{w}|^2 - \bar{\gamma}^2 \sum_{i=k+1}^{\infty} |(\boldsymbol{L}\boldsymbol{w})_i|^2 \\ &\leq \bar{\gamma}^2 \|\boldsymbol{w}\|_2^2 - \bar{\gamma}^2 |L_{k,k}\bar{w}|^2 \end{aligned}$$

Since $\|\boldsymbol{w}\|_2^2 = |\bar{w}|^2 + \sum_{i=k+1}^{\infty} |w_i|^2$, and $\gamma \geq \bar{\gamma}$, we get

$$\begin{aligned} \gamma^2 |\bar{w}|^2 - \gamma^2 |M_{k,k}\bar{w}|^2 &\leq \bar{\gamma}^2 |\bar{w}|^2 + (\bar{\gamma}^2 - \gamma^2) \sum_{i=k+1}^{\infty} |w_i|^2 - \bar{\gamma}^2 |L_{k,k}\bar{w}|^2 \\ &\leq \bar{\gamma}^2 |\bar{w}|^2 - \bar{\gamma}^2 |L_{k,k}\bar{w}|^2 \end{aligned}$$

Since $\bar{w}$ is arbitrary, we can conclude that

$$\gamma^2 I - \gamma^2 M_{k,k}^T M_{k,k} \leq \bar{\gamma}^2 I - \bar{\gamma}^2 L_{k,k}^T L_{k,k}$$

It follows from (4.7) that the entropy decreases as γ increases. ■

4.2.1 Equivalence with the entropy integral

When Definition 4.1 is used to compute the entropy of a linear time-invariant system, the resultant entropy operator coincides with the time-invariant entropy, in the sense that each diagonal element of $\boldsymbol{E}(\boldsymbol{G}, \gamma)$ equals the entropy of the time-invariant system $G(z)$. If $\boldsymbol{G}$ is a causal Toeplitz operator, i.e. the elements of $\boldsymbol{G}$ satisfy $G_{i,j} = g_{i-j}$ for $i \geq j$, then $\boldsymbol{G}$ corresponds to the discrete-time transfer function

$$G(z) = \sum_{k=0}^{\infty} g_k z^{-k}$$

The entropy of the LTI system $G(z)$ follows from the entropy of $\boldsymbol{G}$ as follows:

Lemma 4.8 *Suppose that $\boldsymbol{G} \in \mathcal{C}$ is Toeplitz with $\|\boldsymbol{G}\| < \gamma$. Then*

$$\boldsymbol{E}(\boldsymbol{G}, \gamma) = \operatorname{diag}\{I_d(G, \gamma)\}$$

Proof: Since $\|\boldsymbol{G}\| < \gamma$ there exists a factorization $\boldsymbol{I} - \gamma^{-2}\boldsymbol{G}^*\boldsymbol{G} = \boldsymbol{M}^*\boldsymbol{M}$, where $\boldsymbol{M}$ is a causal Toeplitz operator. It can be checked that

$$I - \gamma^{-2}G(z)^{\sim}G(z) = M(z)^{\sim}M(z)$$

where $M(z) = \sum_{k=0}^{\infty} m_k z^{-k}$. Since $M_{k,k} = m_0 = M(\infty)$ for all k, we can compute

$$\begin{aligned} \boldsymbol{E}(\boldsymbol{G}, \gamma)_k &= -\gamma^2 \ln\det(m_0^T m_0) \\ &= -\gamma^2 \ln\det\left(M(\infty)^T M(\infty)\right) \end{aligned}$$

which equals the entropy of the time-invariant system $G(z)$ by (4.2). ∎

Remark 4.9 If there exists a $p \in \mathbb{N}$ such that $G_{k+p,l+p} = G_{k,l}$ for all $k \geq l$, i.e. $\boldsymbol{G}$ is periodic with period p, it can be shown that there exists a spectral factor $\boldsymbol{M}$ which is also periodic with period p. It is easy to see that the entropy of this operator $\boldsymbol{G}$ will also be periodic with period p.

4.2.2 Entropy in terms of a state-space realization

In Section 3.1 we provided a characterization for the induced operator norm of a causal system represented by a state-space realization. This characterization is in terms of the stabilizing solution to an operator Riccati equation, and this solution can be used to express the entropy in terms of the state-space operators.

Lemma 4.10 *Suppose that $\boldsymbol{G} \in \mathcal{C}$ with $\|\boldsymbol{G}\| < \gamma$ can be represented by a state-space realization (2.1). Then*

$$\boldsymbol{E}(\boldsymbol{G}, \gamma) = -\gamma^2 \operatorname{diag}\left\{\ln\det\left(I - \gamma^{-2}D_k^T D_k - \gamma^{-2}B_k^T X_{k+1} B_k\right)\right\}$$

where $\boldsymbol{0} \leq \boldsymbol{X} \in \mathcal{M}$ is the stabilizing solution to the operator Riccati equation

$$\begin{aligned} \boldsymbol{X} = {} & \boldsymbol{A}^*\boldsymbol{Z}\boldsymbol{X}\boldsymbol{Z}^*\boldsymbol{A} + \boldsymbol{C}^*\boldsymbol{C} \\ & + \gamma^{-2}\left(\boldsymbol{A}^*\boldsymbol{Z}\boldsymbol{X}\boldsymbol{Z}^*\boldsymbol{B} + \boldsymbol{C}^*\boldsymbol{D}\right)\boldsymbol{V}^{-1}\left(\boldsymbol{B}^*\boldsymbol{Z}\boldsymbol{X}\boldsymbol{Z}^*\boldsymbol{A} + \boldsymbol{D}^*\boldsymbol{C}\right) \end{aligned} \tag{4.9}$$

such that $\boldsymbol{V} := \boldsymbol{I} - \gamma^{-2}\boldsymbol{D}^*\boldsymbol{D} - \gamma^{-2}\boldsymbol{B}^*\boldsymbol{Z}\boldsymbol{X}\boldsymbol{Z}^*\boldsymbol{B} > \boldsymbol{0}$.

Proof: This follows immediately from Theorem 3.1, since a spectral factor of $\boldsymbol{I}-\gamma^{-2}\boldsymbol{G}^*\boldsymbol{G}$ is given by

$$\boldsymbol{M}=\boldsymbol{V}^{1/2}-\gamma^{-2}\boldsymbol{V}^{-1/2}(\boldsymbol{B}^*\boldsymbol{Z}\boldsymbol{X}\boldsymbol{Z}^*\boldsymbol{A}+\boldsymbol{D}^*\boldsymbol{C})(\boldsymbol{I}-\boldsymbol{Z}^*\boldsymbol{A})^{-1}\boldsymbol{Z}^*\boldsymbol{B}$$

which has $\boldsymbol{V}^{1/2}$ as the memoryless part. ■

For this expression to be unique it is necessary that $\boldsymbol{X}$ be unique. It is shown in Lemma 3.2 that $\boldsymbol{X}$ is the unique stabilizing solution to (4.9).

Remark 4.11 For a stable LTI system, the elements of the state-space operators are constant matrices A, B, C, and D. It is easily seen that the solution to the operator Riccati equation (4.9) equals $\boldsymbol{X}=\operatorname{diag}\{X\}$, where X is the stabilizing solution to the algebraic Riccati equation in the LTI case. Hence, using the results for the LTI case [53], we have

$$\begin{aligned}\boldsymbol{E}(\boldsymbol{G},\gamma)&=-\gamma^2\operatorname{diag}\left\{\ln\det\left(I-\gamma^{-2}D^TD-\gamma^{-2}B^TXB\right)\right\}\\&=\operatorname{diag}\{I_d(G,\gamma)\}\end{aligned}$$

which shows that the results for the LTI case can be obtained directly using the time-varying entropy, in accordance with Lemma 4.8.

In the next section we discuss the connection between the system theoretic entropy used here and the entropy that is well known in information theory.

4.3 Entropy and information theory

Entropy in system theory is also related to the notion of entropy used in information theory that was introduced by Shannon. In [65] the entropy of a linear time-invariant system is described as the entropy rate of a related system. Here, using a time-domain approach of entropy, we can relate the entropy of an operator to the so-called conditional entropy in information theory. We begin with a brief review of the notion of entropy used in information theory. Standard references are [23, 42].

Consider a random variable $x\in\mathbb{R}^m$. If x can only take a finite number of values, say $x(1),\cdots,x(r)$, then the entropy of x is defined as

$$\mathcal{H}(x):=-\sum_{i=1}^{r}p_i\ln p_i$$

where p_i is the probability that $x = x(i)$. If the random variable is a continuous-type one, this can be extended to

$$\mathcal{H}(x) := -\int_{\mathbb{R}^m} f(x) \ln f(x) \, dx \tag{4.10}$$

where $f(x)$ is the probability density function of x, and by definition we take $f(x) \ln f(x) = 0$ if $f(x) = 0$. Now consider a continuous-type random variable x_k as a function of time. The conditional entropy of order l is defined as

$$\begin{aligned} &\mathcal{H}(x_k | x_{k-1}, \cdots, x_{k-l}) \\ &\quad := -\int_{\mathbb{R}^m} f(x_k | x_{k-1}, \cdots, x_{k-l}) \ln f(x_k | x_{k-1}, \cdots, x_{k-l}) \, dx_k \end{aligned}$$

This is a measure for the uncertainty about its value at time k under the assumption that its l most recent values have been observed. By letting l go to infinity, the conditional entropy of x_k is defined as

$$\mathcal{H}_c(x_k) := \lim_{l \to \infty} \mathcal{H}(x_k | x_{k-1}, \cdots, x_{k-l})$$

assuming the limit exists. The conditional entropy is a measure of the uncertainty about the value of x at time k under the assumption that its entire past is observed.

Now we will look at the entropy in systems theory. It will be shown in Chapter 5 that the notion of entropy defined in this chapter is related to $\mathcal{H}_\infty$, which is associated with the cost function $\|w\|_2^2 - \gamma^{-2} \|z\|_2^2$. This cost is non-negative if the system satisfies the induced norm bound $\|\boldsymbol{G}\| < \gamma$. Since $\boldsymbol{I} - \gamma^{-2} \boldsymbol{G}^* \boldsymbol{G} = \boldsymbol{M}^* \boldsymbol{M}$, the operator $\boldsymbol{M}$ is a linear operator mapping $\boldsymbol{w}$ to a signal related to the cost function used in $\mathcal{H}_\infty$ theory. This linearity assures that, if $\boldsymbol{w}$ is a normal process sequence, the output $\boldsymbol{M}\boldsymbol{w}$ is also a normal process sequence. The causality of $\boldsymbol{M}$ is essential in relating our entropy to the conditional entropy in information theory.

Lemma 4.12 $\mathcal{H}_c((\boldsymbol{M}\boldsymbol{w})_k) = \mathcal{H}_c(w_k) + \frac{1}{2} \ln \det(M_{k,k}^T M_{k,k})$

Proof: We are interested in the conditional entropy of $(\boldsymbol{M}\boldsymbol{w})_k$. Since $\boldsymbol{M}$ is causal with a causal inverse, it is easy to see that $(\boldsymbol{M}\boldsymbol{w})_i$ is known for all $i < k$ if and only if w_i is known for all $i < k$. Thus, we can evaluate

$$\begin{aligned} \mathcal{H}_c((\boldsymbol{M}\boldsymbol{w})_k) &= \lim_{l \to \infty} \mathcal{H}((\boldsymbol{M}\boldsymbol{w})_k | (\boldsymbol{M}\boldsymbol{w})_{k-1}, \cdots, (\boldsymbol{M}\boldsymbol{w})_{k-l}) \\ &= \lim_{l \to \infty} \mathcal{H}((\boldsymbol{M}\boldsymbol{w})_k | w_{k-1}, \cdots, w_{k-l}) \end{aligned}$$

For a Gaussian random variable $x \in \mathbb{R}^m$, define $v := x - \mathcal{E}x$, and also define $R := \mathcal{E}vv^T \geq 0$. If $R > 0$, the probability density function equals

$$f(x) = \frac{|\det R|^{-1/2}}{\sqrt{(2\pi)^m}} \exp\left(-\frac{1}{2}v^T R^{-1} v\right)$$

Hence, with $r := R^{-1/2}v$, the entropy (4.10) of x can be computed using $dv = |\det R|^{1/2}\, dr$ as

$$\begin{aligned}
\mathcal{H}(x) &= -\int_{\mathbb{R}^m} f(x) \ln f(x)\, dx \\
&= -\int_{\mathbb{R}^m} \frac{1}{\sqrt{(2\pi)^m}} \exp\left(-\frac{1}{2}r^T r\right) \ln |\det R|^{-1/2}\, dr \\
&\quad - \int_{\mathbb{R}^m} \frac{1}{\sqrt{(2\pi)^m}} \exp\left(-\frac{1}{2}r^T r\right) \ln \frac{1}{\sqrt{(2\pi)^m}} \exp\left(-\frac{1}{2}r^T r\right) dr \\
&= \frac{1}{2} \ln |\det R| + \mathcal{H}(r) \qquad (4.11)
\end{aligned}$$

Since $(\boldsymbol{M}\boldsymbol{w})_k = \sum_{i=-\infty}^{k} M_{k,i} w_i$, it is easy to see that

$$\mathcal{E}\left\{(\boldsymbol{M}\boldsymbol{w})_k \Big| w_i \text{ for all } i < k\right\} = M_{k,k}\mathcal{E}w_k + \sum_{i=-\infty}^{k-1} M_{k,i} w_i$$

Furthermore the variance can be computed as

$$\begin{aligned}
&\mathcal{E}\left\{((\boldsymbol{M}\boldsymbol{w})_k - \mathcal{E}(\boldsymbol{M}\boldsymbol{w})_k)\,((\boldsymbol{M}\boldsymbol{w})_k - \mathcal{E}(\boldsymbol{M}\boldsymbol{w})_k)^T \Big| w_i \text{ for all } i < k\right\} \\
&\qquad = M_{k,k}\mathcal{E}(w_k - \mathcal{E}w_k)(w_k - \mathcal{E}w_k)^T M_{k,k}^T
\end{aligned}$$

Knowing the expectation and the variance of the distribution, the result follows from an evaluation of the entropy as in (4.11). ■

The difference between the two conditional entropies equals (up to a factor $-2\gamma^2$) the entropy that has been introduced here for time-varying systems. Since the signal w_k is the input signal, its conditional entropy does not depend on the system. The entropy of $(\boldsymbol{M}\boldsymbol{w})_k$ however, *does* depend on the system, and can be changed by choosing a different controller. Clearly, minimizing the system's entropy assures that, in case the past of w_k is known, observing w_k gains more information about the cost function $\|\boldsymbol{w}\|_2^2 - \gamma^{-2}\|\boldsymbol{z}\|_2^2$ than other controllers achieve.

Remark 4.13 The entropy rate of a process is defined in terms of the joint entropy as

$$\bar{\mathcal{H}}(x) = \lim_{l\to\infty} \frac{1}{2l+1}\mathcal{H}(x_{-l}, \cdots, x_l)$$

which is an average entropy. In [65] it is shown that the entropy of an LTI system is related to the difference of the entropy rates of w and Mw. The conditional entropy equals the entropy rate at every time k if the system is time-invariant, but this is obviously not the case for time-varying systems.

In the next section we will define the notion of entropy for an anti-causal operator, and discuss the connection with the entropy of its causal adjoint.

4.4 Entropy of an anti-causal system

Definition 4.1 provides a suitable definition of the entropy for a causal system. In the LTI case the entropy of a causal system equals the entropy of its (anti-causal) adjoint [64]. This is easy to see, since, inside the integral in (4.1)

$$\det\left(I - \gamma^{-2}G(e^{-i\omega})^T G(e^{i\omega})\right) = \det\left(I - \gamma^{-2}G(e^{i\omega})G(e^{-i\omega})^T\right)$$

For time-varying systems, we must reformulate the notion of entropy starting by reexamining the spectral factorization used. We now introduce a suitable definition of entropy for an anti-causal operator $\boldsymbol{H}$. Recall that $\boldsymbol{H}$ is anti-causal iff $\boldsymbol{H}^*$ is causal.

Definition 4.14 *Suppose that $\boldsymbol{H}^* \in \mathcal{C}$ and $\|\boldsymbol{H}\| < \gamma$. Let $\boldsymbol{N} \in \mathcal{C} \cap \mathcal{C}^{-1}$ be a co-spectral factor of the positive operator $\boldsymbol{I} - \gamma^{-2}\boldsymbol{H}^*\boldsymbol{H}$, which exists according to Lemma 2.4. We define the entropy by*

$$\boldsymbol{E}_a(\boldsymbol{H}, \gamma) := -\gamma^2 \operatorname{diag}\left\{\ln\det\left(N_{k,k}N_{k,k}^T\right)\right\} \in \mathcal{M}$$

The subscript "a" in the definition indicates that the argument $\boldsymbol{H}$ is an anti-causal operator. The entropy for an anti-causal system satisfies properties similar to Lemma 4.5 and Lemma 4.10 for causal systems. These we will state here without proof.

Corollary 4.15 *Suppose that $\boldsymbol{H}^* \in \mathcal{C}$ with $\|\boldsymbol{H}\| < \gamma$. Then $\boldsymbol{0} \le \boldsymbol{E}_a(\boldsymbol{H}, \gamma)$ with equality if and only if $\boldsymbol{H} = \boldsymbol{0}$.* □

Corollary 4.16 *Suppose that $\boldsymbol{H}^* \in \mathcal{C}$ with $\|\boldsymbol{H}\| < \gamma$ can be represented by a state-space realization (2.1). Then*

$$\boldsymbol{E}_a(\boldsymbol{H},\gamma) = -\gamma^2 \operatorname{diag}\left\{\ln\det\left(I - \gamma^{-2}D_kD_k^T - \gamma^{-2}C_kY_{k-1}C_k^T\right)\right\}$$

where $\mathbf{0} \leq \boldsymbol{Y} \in \mathcal{M}$ is the stabilizing solution to the operator Riccati equation

$$\begin{aligned}\boldsymbol{Y} &= \boldsymbol{A}\boldsymbol{Z}^*\boldsymbol{Y}\boldsymbol{Z}\boldsymbol{A}^* + \boldsymbol{B}\boldsymbol{B}^* \\ &\quad + \gamma^{-2}\left(\boldsymbol{A}\boldsymbol{Z}^*\boldsymbol{Y}\boldsymbol{Z}\boldsymbol{C}^* + \boldsymbol{B}\boldsymbol{D}^*\right)\boldsymbol{U}^{-1}\left(\boldsymbol{C}\boldsymbol{Z}^*\boldsymbol{Y}\boldsymbol{Z}\boldsymbol{A}^* + \boldsymbol{D}\boldsymbol{B}^*\right)\end{aligned}$$

such that $\boldsymbol{U} := \boldsymbol{I} - \gamma^{-2}\boldsymbol{D}\boldsymbol{D}^* - \gamma^{-2}\boldsymbol{C}\boldsymbol{Z}^*\boldsymbol{Y}\boldsymbol{Z}\boldsymbol{C}^* > \mathbf{0}$. □

The following example shows that $\boldsymbol{E}(\boldsymbol{G},\gamma) \neq \boldsymbol{E}_a(\boldsymbol{G}^*,\gamma)$ in general.

Example 4.17 Let $\boldsymbol{G}$ be given by

$$\boldsymbol{G} = \begin{bmatrix} \mathbf{0} & \mathbf{0} & \mathbf{0} & \mathbf{0} \\ \mathbf{0} & \boxed{\frac{1}{2}} & 0 & \mathbf{0} \\ \mathbf{0} & \frac{1}{2} & 0 & \mathbf{0} \\ \mathbf{0} & \mathbf{0} & \mathbf{0} & \mathbf{0} \end{bmatrix}$$

where the box denotes the $(0,0)$ element. It is easily checked that this operator corresponds to a state-space realization $(\boldsymbol{A},\boldsymbol{B},\boldsymbol{C},\boldsymbol{D})$ with matrices $B_0 = D_0 = \frac{1}{2}$, $C_1 = 1$, and all the other elements are zero. We can compute (take $\boldsymbol{H} = \boldsymbol{G}^*$)

$$\begin{aligned}\boldsymbol{I} - \boldsymbol{G}^*\boldsymbol{G} &= \begin{bmatrix} \boldsymbol{I} & \mathbf{0} & \mathbf{0} & \mathbf{0} \\ \mathbf{0} & \boxed{\frac{1}{2}} & 0 & \mathbf{0} \\ \mathbf{0} & 0 & 1 & \mathbf{0} \\ \mathbf{0} & \mathbf{0} & \mathbf{0} & \boldsymbol{I} \end{bmatrix} \\ &= \begin{bmatrix} \boldsymbol{I} & \mathbf{0} & \mathbf{0} & \mathbf{0} \\ \mathbf{0} & \boxed{\sqrt{\frac{1}{2}}} & 0 & \mathbf{0} \\ \mathbf{0} & 0 & 1 & \mathbf{0} \\ \mathbf{0} & \mathbf{0} & \mathbf{0} & \boldsymbol{I} \end{bmatrix} \begin{bmatrix} \boldsymbol{I} & \mathbf{0} & \mathbf{0} & \mathbf{0} \\ \mathbf{0} & \boxed{\sqrt{\frac{1}{2}}} & 0 & \mathbf{0} \\ \mathbf{0} & 0 & 1 & \mathbf{0} \\ \mathbf{0} & \mathbf{0} & \mathbf{0} & \boldsymbol{I} \end{bmatrix}\end{aligned}$$

Similarly, for the co-spectral factor

$$
I - GG^* = \begin{bmatrix} I & 0 & 0 & 0 \\ 0 & \boxed{\frac{3}{4}} & -\frac{1}{4} & 0 \\ 0 & -\frac{1}{4} & \frac{3}{4} & 0 \\ 0 & 0 & 0 & I \end{bmatrix}
$$

$$
= \begin{bmatrix} I & 0 & 0 & 0 \\ 0 & \boxed{\sqrt{\frac{3}{4}}} & 0 & 0 \\ 0 & -\sqrt{\frac{1}{12}} & \sqrt{\frac{2}{3}} & 0 \\ 0 & 0 & 0 & I \end{bmatrix} \begin{bmatrix} I & 0 & 0 & 0 \\ 0 & \boxed{\sqrt{\frac{3}{4}}} & -\sqrt{\frac{1}{12}} & 0 \\ 0 & 0 & \sqrt{\frac{2}{3}} & 0 \\ 0 & 0 & 0 & I \end{bmatrix}
$$

The entropies are given by

$$
E(G) = \begin{bmatrix} 0 & 0 & 0 & 0 \\ 0 & \boxed{\ln 2} & 0 & 0 \\ 0 & 0 & 0 & 0 \\ 0 & 0 & 0 & 0 \end{bmatrix}
$$

and

$$
E_a(G^*) = \begin{bmatrix} 0 & 0 & 0 & 0 \\ 0 & \boxed{\ln \frac{4}{3}} & 0 & 0 \\ 0 & 0 & \ln \frac{3}{2} & 0 \\ 0 & 0 & 0 & 0 \end{bmatrix}
$$

which are obviously not equal. □

Though not equal, the two entropies are related to each other, through the time-reverse operator Ω that was introduced in Section 2.3. It is shown in Corollary 2.11 that $\Omega G \Omega$ is an anti-causal operator, and its induced operator norm equals that of G. The entropies of these systems are related as follows:

Lemma 4.18 *Suppose that $G \in \mathcal{C}$ with $\|G\| < \gamma$. Then*

$$
E(G, \gamma) = \Omega E_a(\Omega G \Omega, \gamma) \Omega
$$

Proof: Suppose that $\boldsymbol{M}$ is a spectral factor of $\boldsymbol{I} - \gamma^{-2}\boldsymbol{G}^*\boldsymbol{G}$. Then, recalling the facts that $\Omega^2 = \boldsymbol{I}$ and that $\Omega^* = \Omega$

$$\begin{aligned}
\boldsymbol{I} - \gamma^{-2}(\Omega\boldsymbol{G}\Omega)^*(\Omega\boldsymbol{G}\Omega) &= \boldsymbol{I} - \gamma^{-2}\Omega\boldsymbol{G}^*\boldsymbol{G}\Omega \\
&= \Omega(\boldsymbol{I} - \gamma^{-2}\boldsymbol{G}^*\boldsymbol{G})\Omega \\
&= \Omega\boldsymbol{M}^*\boldsymbol{M}\Omega \\
&= (\Omega\boldsymbol{M}^*\Omega)(\Omega\boldsymbol{M}^*\Omega)^*
\end{aligned}$$

Since $\boldsymbol{M}$ is causal, we also have that $\Omega\boldsymbol{M}^*\Omega$ is causal. Hence $\Omega\boldsymbol{M}^*\Omega$ is a co-spectral factor of $\boldsymbol{I} - \gamma^{-2}(\Omega\boldsymbol{G}\Omega)^*(\Omega\boldsymbol{G}\Omega)$. Thus

$$\begin{aligned}
\boldsymbol{E}_a(\Omega\boldsymbol{G}\Omega, \gamma) &= -\gamma^2 \operatorname{diag}\left\{\ln\det\left((\Omega\boldsymbol{M}^*\Omega)_{k,k}(\Omega\boldsymbol{M}^*\Omega)_{k,k}^T\right)\right\} \\
&= -\gamma^2 \operatorname{diag}\left\{\ln\det\left(M_{-k,-k}^T M_{-k,-k}\right)\right\} \\
&= -\gamma^2\Omega \operatorname{diag}\left\{\ln\det\left(M_{k,k}^T M_{k,k}\right)\right\}\Omega \\
&= \Omega\boldsymbol{E}(\boldsymbol{G}, \gamma)\Omega
\end{aligned}$$

and the result follows. ■

If the system is time-invariant (i.e. $\boldsymbol{G}$ is a Toeplitz operator), the spectral factor is also Toeplitz. Hence, every diagonal element of $\boldsymbol{E}(\boldsymbol{G}, \gamma)$ is the same and equal to the entropy of the corresponding LTI system $G(z)$. It follows that

$$\boldsymbol{E}(\boldsymbol{G}, \gamma) = \boldsymbol{E}_a(\Omega\boldsymbol{G}\Omega, \gamma) \tag{4.12}$$

Since $\Omega\boldsymbol{G}\Omega$ is the operator $\boldsymbol{G}$ with time-reversed inputs and outputs, equation (4.12) corresponds to the fact that the entropy of an LTI system equals the entropy of its adjoint. This can be shown directly.

Lemma 4.19 *Suppose that $\boldsymbol{G} \in \mathcal{C}$ is Toeplitz with $\|\boldsymbol{G}\| < \gamma$. Then $\boldsymbol{E}(\boldsymbol{G}, \gamma) = \boldsymbol{E}_a(\boldsymbol{G}^*, \gamma)$.*

Proof: If $\boldsymbol{G}$ is causal and Toeplitz, say $G_{i,j} = g_{i-j}$ for $i \geq j$, then $\boldsymbol{G}$ corresponds to the time-invariant transfer function

$$G(z) = \sum_{k=0}^{\infty} g_k z^{-k}$$

Since $\|\boldsymbol{G}\| < \gamma$ there exist spectral and co-spectral factorizations

$$\boldsymbol{I} - \gamma^{-2}\boldsymbol{G}^*\boldsymbol{G} = \boldsymbol{M}^*\boldsymbol{M} \text{ and } \boldsymbol{I} - \gamma^{-2}\boldsymbol{G}\boldsymbol{G}^* = \boldsymbol{N}\boldsymbol{N}^*$$

where $\boldsymbol{M}$ and $\boldsymbol{N}$ are both causal Toeplitz operators. In Lemma 4.8 we showed that

$$\boldsymbol{E}(\boldsymbol{G},\gamma) = -\frac{\gamma^2}{2\pi}\,\text{diag}\left\{\int_{-\pi}^{\pi} \ln\det\left(I-\gamma^{-2}G(e^{-i\omega})^T G(e^{i\omega})\right)\,d\omega\right\}$$

Similarly, we find that

$$\boldsymbol{E}_a(\boldsymbol{G}^*,\gamma) = -\frac{\gamma^2}{2\pi}\,\text{diag}\left\{\int_{-\pi}^{\pi} \ln\det\left(I-\gamma^{-2}G(e^{i\omega})G(e^{-i\omega})^T\right)\,d\omega\right\}$$

The results follows by elementary property of determinants. ∎

4.5 Entropy and the $\mathcal{W}$-transform

For linear time-invariant systems, generalizations of the entropy integral are known that allow one to compute the entropy at different points z_0 outside the unit circle [6, 34, 54]. In particular, the entropy, evaluated at the point z_0, $|z_0| > 1$ is given by the integral

$$I_d(G,\gamma,z_0) := -\frac{\gamma^2}{2\pi}\int_{-\pi}^{\pi} \ln\left|\det\left(I-\gamma^{-2}G(e^{-i\omega})^T G(e^{i\omega})\right)\right| \frac{|z_0|^2-1}{|z_0-e^{i\omega}|^2}d\omega$$

A similar expression is known for continuous-time systems [32, 64]. The formula for the entropy given above is equivalent to the evaluation of the spectral factor $M(z)$ at the desired point z_0. For time-varying systems, the entropy introduced here has a similar generalization in terms of the $\mathcal{W}$-transform. This transform was introduced in [1] in order to consider interpolation problems for non-Toeplitz operators. This transform generalizes the usual $\mathcal{Z}$-transform to time-varying systems.

Given an operator $\boldsymbol{G} \in \mathcal{C}$, $\boldsymbol{G} = \{G_{i,j}\}$, define the set of diagonal operators $\boldsymbol{G}_{[k]}$ corresponding to the kth subdiagonal shifted up to the diagonal:

$$\boldsymbol{G}_{[k]} = \text{diag}\{\ldots, G_{-1,-1-k}, \boxed{G_{0,-k}}, G_{1,1-k},\ldots\}$$

From Lemmas 2.1 and 2.2,

$$\left\|\boldsymbol{G}_{[k]}\right\| = \sup_i \|G_{i,i-k}\| \leq \|\boldsymbol{G}\| \tag{4.13}$$

An operator $\boldsymbol{G} \in \mathcal{C}$ has a unique representation as a series in terms of the $\boldsymbol{G}_{[k]}$ as follows

$$\boldsymbol{G} \sim \sum_{k=0}^{\infty} \boldsymbol{G}_{[k]}(\boldsymbol{Z}^*)^k$$

For a compact $\boldsymbol{G}$, the sum converges in norm to $\boldsymbol{G}$ and we can write an equality. For a non-compact $\boldsymbol{G}$, the sum need not converge in the induced norm. For this class of systems, we can think of this expression in the following sense. The difference between $\boldsymbol{G}$ and the nth partial sum as belonging to the space $(\boldsymbol{Z}^*)^n\mathcal{C}$ and the $\boldsymbol{G}_{[k]}$ is the unique set of diagonal operators $\boldsymbol{D}_k$ for which

$$\boldsymbol{G} - \sum_{k=0}^{n} \boldsymbol{D}_k(\boldsymbol{Z}^*)^k \in (\boldsymbol{Z}^*)^n\mathcal{C}$$

Let $\boldsymbol{W} \in \mathcal{M}$, and define

$$\ell(\boldsymbol{W}) := \rho(\boldsymbol{W}\boldsymbol{Z}^*)$$

We are now in a position to define the $\mathcal{W}$-transform.

Definition 4.20 *Let $\boldsymbol{G} \in \mathcal{C}$ and $\boldsymbol{W} \in \mathcal{M}$, with $\ell(\boldsymbol{W}) < 1$. We define*

$$\widehat{\boldsymbol{G}}(\boldsymbol{W}) := \sum_{k=0}^{\infty} \boldsymbol{G}_{[k]}(\boldsymbol{Z}^*)^k(\boldsymbol{Z}\boldsymbol{W})^k \tag{4.14}$$

This series will converge in norm provided $\ell(\boldsymbol{W}) < 1$. The transform (4.14) acts like the λ-transform in the time-invariant case.[1]

The following two examples are given to clarify the notion of the $\mathcal{W}$-transform. We first apply the definition of the transform to an LTI system.

Example 4.21 (Time-Invariant Systems) For an LTI system $\boldsymbol{G} \in \mathcal{C}$, the block Toeplitz matrix has diagonal elements $G_{i,i-k} = g_k$. It follows that

$$\boldsymbol{G}_{[k]} = \text{diag}\{\ldots, g_k, \boxed{g_k}, g_k, \ldots\}$$

Let $\boldsymbol{W} = \lambda \boldsymbol{I}$, where $\lambda \in \mathbb{C}$, $|\lambda| < 1$ and $\boldsymbol{I}$ is the identity operator in $\mathcal{B}$. Then

$$\widehat{\boldsymbol{G}}(\boldsymbol{W}) = \sum_{k=0}^{\infty} \boldsymbol{G}_{[k]}(\boldsymbol{Z}^*)^k(\lambda\boldsymbol{Z})^k = \sum_{k=0}^{\infty} \lambda^k \boldsymbol{G}_{[k]} = \text{diag}\{G(\lambda)\}$$

where $G(\lambda)$ is the usual λ-transform of the sequence $\{g_k\}$. □

In the second example, we again compute the $\mathcal{W}$-transform at the point $\boldsymbol{W} = \lambda\boldsymbol{I}$, but we do so for a general time-varying system

[1] The λ-transform is just the $\mathcal{Z}$-transform with z^{-1} replaced by λ.

Example 4.22 (Time-Varying Systems) For an LTV system $\boldsymbol{G} \in \mathcal{C}$, we can compute

$$\widehat{\boldsymbol{G}}(\lambda \boldsymbol{I}) = \text{diag}\{\dots, G_{-1}(\lambda), \boxed{G_0(\lambda)}, G_1(\lambda), \dots\}$$

where

$$G_i(\lambda) = \sum_{k=0}^{\infty} G_{i,i-k}\lambda^k, \qquad i \in \mathbb{Z}$$

is the λ-transform of the frozen-time system at time i. These frozen-time systems have received considerable interest recently in the study of slowly-time varying systems [79, 80, 86]. If we let $\lambda = e^{-i\omega}$, then we recover Zadeh's system function [83]. Note that in this case we must prove convergence of the sum defining the transform in some other way. □

This last example demonstrates that the $\mathcal{W}$-transform includes one of the better known frequency domain functions for time-varying systems.

The $\mathcal{W}$-transform allows us to generalize the definition of the entropy.

Definition 4.23 Suppose that $\boldsymbol{G} \in \mathcal{C}$ and $\|\boldsymbol{G}\| < \gamma$. Let $\boldsymbol{M} \in \mathcal{C} \cap \mathcal{C}^{-1}$ be a spectral factor of the positive operator $\boldsymbol{I} - \gamma^{-2}\boldsymbol{G}^*\boldsymbol{G}$ and let $\boldsymbol{W} \in \mathcal{M}$ be a diagonal operator with $\ell(\boldsymbol{W}) < 1$. We define the entropy evaluated at $\boldsymbol{W}$ as the operator:

$$\boldsymbol{E}(\boldsymbol{G}, \gamma, \boldsymbol{W}) := \widehat{\boldsymbol{M}}^*(\boldsymbol{W})\widehat{\boldsymbol{M}}(\boldsymbol{W}) \tag{4.15}$$

Remark 4.24 As was done for the entropy of Definition 4.1, it is straightforward to check that the entropy of Definition 4.23 does not depend on the spectral factor chosen.

For engineering applications, the most desirable location for the evaluation of the entropy is $z_0 = \infty$ which corresponds to $\boldsymbol{W} = \boldsymbol{0}$. At this point, Definitions 4.1 and 4.23 coincide. For this reason we will not make further use of the generalized entropy definition.

For time-invariant systems, the generalization of the entropy presented here is useful when dealing with minimum entropy control problems via the bilinear transformation $s = \frac{z+1}{z-1}$; see [49]. For a discrete-time transfer function $G(z)$, we can define a continuous-time transfer function

$$\tilde{G}(s) = G\left(\frac{1+s}{1-s}\right)$$

and compute the entropy. It is well known that the $\mathcal{H}_\infty$ norm of the transfer function is invariant under this transformation, i.e. the discrete-time

$\mathcal{H}_\infty$ norm of G equals the continuous-time $\mathcal{H}_\infty$ norm of $\tilde{G}$. Moreover, the entropies will also be the same, in the sense that

$$I_c(\tilde{G}, \gamma, \tfrac{z_0+1}{z_0-1}) = I_d(G, \gamma, z_0)$$

for the appropriately defined I_c; see [64]. In order to compute the discrete-time entropy at the desired point ($z_0 = \infty$), however, the continuous-time entropy must be computed at $s_0 = 1$, but this is not the ideal point $s_0 = \infty$. For time-varying systems it is not known whether there exists an analogous relationship. In Chapter 7 an entropy formula for continuous-time, time-varying systems is presented which is analogous to the discrete-time Definition 4.1.

4.6 Entropy of a non-linear system

We have extended the notion of entropy to the class of linear time-varying systems. Since entropy has some very attractive features, it is of interest to consider extensions to other classes of systems. Our entropy is defined in terms of a spectral factorization, which other classes of systems also admit. For example, for continuous-time non-linear systems, the spectral factorization problem was considered in [10]. It is of interest to investigate whether this factorization can be used to define the entropy for non-linear systems. We will outline this for systems described by the non-linear differential equation

$$\Sigma_G := \left\{ \begin{array}{rcl} \dot{x}_t & = & A(x_t) + B(x_t)w_t \\ z_t & = & C(x_t) \end{array} \right. \tag{4.16}$$

where each of the $A(x)$, $B(x)$, and $C(x)$ are smooth functions. The vectors $x \in \mathbb{R}^n$, $w \in \mathbb{R}^m$ and $z \in \mathbb{R}^p$ are the state, input, and output respectively. We assume the existence of a globally exponentially asymptotically stable equilibrium, i.e. $\bar{x}$ is such that $A(\bar{x}) = 0$, and without loss of generality, we assume that $C(\bar{x}) = 0$.

We can define, along the lines of [77], the $\mathcal{L}_{2+}$ gain of the system.

Definition 4.25 *A system G is said to have an $\mathcal{L}_{2+}$ gain less than or equal to γ if*

$$\lim_{T \to \infty} \sup_{w \in \mathcal{L}_2[0,T]} \int_0^T (\|z_t\|_2^2 - \gamma^2 \|w_t\|_2^2)\, dt \le 0$$

where $x_0 = \bar{x}$.

The following result provides a sufficient condition that guarantees that the $\mathcal{L}_{2+}$ gain of a system is bounded by γ.

Lemma 4.26 ([77]) *If there exists a smooth function $V(x) \geq 0$ that satisfies the Hamilton-Jacobi equation*

$$\frac{\partial V}{\partial x}(x)A(x) + \frac{1}{2\gamma^2}\frac{\partial V}{\partial x}(x)B(x)B^T(x)\left(\frac{\partial^T V}{\partial x}(x)\right) + \frac{1}{2}C^T(x)C(x) = 0 \tag{4.17}$$

with $V(\bar{x}) = 0$, then the system (4.16) has an $\mathcal{L}_{2+}$ gain less than or equal to γ. □

Remark 4.27 It is shown in [77] that the converse is also true if the system is reachable from $\bar{x}$.

For the input-output mapping associated with a system, we say that $T : \mathcal{L}_{2+}^m \to \mathcal{L}_{2+}^m$ is outer if T has a finite $\mathcal{L}_{2+}$ gain, and the inverse map of T exists and also has a finite $\mathcal{L}_{2+}$ gain.

We will also require the notion of the *Fréchet* derivative of a mapping $T : \mathcal{L}_{2+}^m \to \mathcal{L}_{2+}^p$ at a point $w \in \mathcal{L}_{2+}^m$, denoted by $DT(w)$; see [71]. This is a linear mapping from $\mathcal{L}_{2+}^m$ to $\mathcal{L}_{2+}^p$ and so admits a transpose.

Now, assuming that (4.16) has $\mathcal{L}_{2+}$ gain less than or equal to γ, we are interested in a spectral factorization. Using Lemma 4.26, a spectral factorization can be defined as in [10].

Lemma 4.28 *Under the assumptions of Lemma 4.26, there exists an outer system M, such that*

$$I - \gamma^{-2}[DG]^T \circ G(w) = [DM]^T \circ M(w)$$

Moreover

$$\Sigma_M := \begin{cases} \dot{x}_t &= A(x_t) + B(x_t)w_t \\ z_t &= -\gamma^{-2}B^T(x_t)\dfrac{\partial^T V}{\partial x}(x_t) + w_t \end{cases}$$

is a state-space realization of such a system. □

Since $V(x)$ is a smooth function with $V(\bar{x}) = 0$, we can use Morse's lemma (see [62]) to write

$$V(x) = (x - \bar{x})^T H(x)(x - \bar{x})$$

The entropy of the system G can be defined in terms of this matrix $H(x)$.

Definition 4.29 *The entropy of a system G is defined as*

$$E(G,\gamma) := \mathrm{tr}\left[B(\bar{x})^T H(\bar{x}) B(\bar{x})\right]$$

Remark 4.30 It is easily checked that, at $x = \bar{x}$, we have

$$H(\bar{x}) = \frac{1}{2}\frac{\partial^2 V}{\partial x^2}(\bar{x}) \tag{4.18}$$

which gives an expression for the entropy in terms of $V(x)$.

Remark 4.31 If the system is linear, i.e. $A(x) = Ax$, $B(x) = B$ and $C(x) = Cx$, then the equilibrium point is $\bar{x} = 0$, and the solution to the Hamilton-Jacobi equation (4.17) equals $x^T X x$, where X is the stabilizing positive semi-definite solution to the algebraic Riccati equation

$$A^T X + XA + \frac{1}{\gamma^2} XBB^T X + C^T C = 0$$

Hence the entropy equals $\mathrm{tr}\left[B^T X B\right]$, which equals the state-space expression for the entropy of a continuous-time, linear time-invariant system.

Example 4.32 Consider the system from [61], given by

$$\begin{aligned} \dot{x}_1 &= x_2 \\ \dot{x}_2 &= -x_1^3 - x_2 + w \\ z &= x_2 \end{aligned}$$

Obviously, the equilibrium point equals $\bar{x} = 0$. For $\gamma \geq 1$ the Hamilton-Jacobi equation (4.17) has a solution $V(x) = \frac{1}{4}\alpha(x_1^4 + 2x_2^2)$, where

$$\alpha = \gamma^2 - \gamma\sqrt{\gamma^2 - 1}$$

Now,

$$V(x) = \begin{bmatrix} x_1 & x_2 \end{bmatrix} \begin{bmatrix} \frac{\alpha}{4}x_1^2 & 0 \\ 0 & \frac{\alpha}{2} \end{bmatrix} \begin{bmatrix} x_1 \\ x_2 \end{bmatrix}$$

and so the entropy equals $\frac{\alpha}{2}$. □

Because Definition (4.29) has been given in terms of a state-space realization, it is necessary to show that the entropy does not depend on the particular coordinates. Recall (eg. [57]) that a non-linear change of coordinates involves a transformation

$$v = \phi(x)$$

where $\phi(x)$ is a global diffeomorphism on $\mathbb{R}^n$; i.e. $\phi(x)$ is invertible (for all $x \in \mathbb{R}^n$) and both ϕ and ϕ^{-1} are smooth mappings. Using the new state vector v, the system (4.16) can be expressed as

$$\begin{aligned} \dot{v}_t &= \bar{A}(v_t) + \bar{B}(v_t)w_t \\ z_t &= \bar{C}(v_t) \end{aligned}$$

with equilibrium point $\bar{v} = \phi(\bar{x})$. Here

$$\begin{aligned} \bar{A}(v) &= \left[\frac{\partial \phi}{\partial x}(x)A(x)\right]_{x=\phi^{-1}(v)} \\ \bar{B}(v) &= \left[\frac{\partial \phi}{\partial x}(x)B(x)\right]_{x=\phi^{-1}(v)} \end{aligned}$$

and $\bar{C}(v) = C(\phi^{-1}(v))$. It is straightforward to show that the function $\bar{V}$ given by $\bar{V}(v) := V(x)|_{x=\phi^{-1}(v)}$ will satisfy the Hamilton-Jacobi equality for the new coordinates. Moreover

$$\frac{\partial^2 V}{\partial x^2} = \left(\frac{\partial \phi}{\partial x}\right)^T \frac{\partial^2 \bar{V}}{\partial v^2} \left(\frac{\partial \phi}{\partial x}\right)$$

Using (4.18), it is easy to compute the entropy (evaluated at $\bar{v} = \phi(\bar{x})$) as

$$\frac{1}{2}\operatorname{tr}\left[\bar{B}^T(\bar{v})\frac{\partial^2 \bar{V}}{\partial v^2}(\bar{v})\bar{B}(\bar{v})\right] = \frac{1}{2}\operatorname{tr}\left[B^T(\bar{x})\frac{\partial^2 V}{\partial x^2}(\bar{x})B(\bar{x})\right]$$

It follows that the entropy does not depend on the particular coordinates chosen.

The derivations presented here can provide a starting point for the study of the minimum entropy control problem for non-linear systems. One of the first steps must be to ensure that Definition 4.29 makes sense, in that the entropy defined is a property of the system. This will require several steps. First of all, we have defined the entropy in terms of the function $H(x)$ which arises from the solution to the Hamilton-Jacobi equation. Note that this equation will in general have more than one solution. For linear systems we take the stabilizing solution of the Riccati equation. What the correct $V(x)$ in the non-linear case remains to this point unclear.

A second potential for non-uniqueness arises from the choice of spectral factor. For linear systems, it was shown that the entropy did not depend on the spectral factor chosen. This was done by using the fact that all spectral factors are related by a memoryless, unitary operator. Whether this property exists for non-linear systems needs to be investigated.

Finally, we mention that the entropy optimization problem for linear systems was solved by using the property that all solutions to the $\mathcal{H}_\infty$ sub-optimal control problem can be parameterized as a linear fractional transformation of an inner system and an arbitrary contraction. Whether the same is true for non-linear systems remains an open problem.

Connections With Related Optimal Control Problems 5

In this chapter we consider the connection between the entropy and some related optimal control problems. In particular, it will be shown that the relationship that exists between the entropy of a time-invariant system and the $\mathcal{H}_2$ norm also carries over to the entropy formulation for time-varying systems that was introduced in Chapter 4. Specifically, it will be shown that the entropy is an upper bound for a quadratic cost operator, while maintaining a bound on the $\mathcal{H}_\infty$ norm. It follows that the minimum entropy controller allows the designer to trade off between the robustness of the system and its $\mathcal{H}_2$ performance.

Having established this connection, we then consider the relationship between the average entropy of a causal system and the average entropy of its anti-causal adjoint.

Finally, the connection with risk-sensitive control is investigated. For time-invariant systems, the entropy functional and the linear exponential quadratic Gaussian cost are the same. In this chapter, it is shown that this is not true for general time-varying systems. It does hold, however, when the system admits a state-space representation. It also holds for systems defined over a finite time interval, a result which is new even for linear time-invariant systems.

5.1 Relationship with $\mathcal{H}_\infty$ control

As was shown in Chapter 1, the $\mathcal{H}_\infty$ control problem is related to the cost function

$$J(\boldsymbol{w},\gamma) := \|\boldsymbol{w}\|_2^2 - \gamma^{-2}\|\boldsymbol{z}\|_2^2$$

where $\boldsymbol{z} = \boldsymbol{G}\boldsymbol{w}$. For a causal operator $\boldsymbol{G}$ with $\|\boldsymbol{G}\| < \gamma$, it is easy to see that $J(\boldsymbol{w},\gamma)$ is non-negative for all $\boldsymbol{w} \in \ell_2$, with equality if and only if $\boldsymbol{w} = \boldsymbol{0}$. The input $\boldsymbol{w} = \boldsymbol{0}$ is therefore known as the worst-case input.

For a causal system, whenever the input w_i is known for all $i < k$ at time k, the worst-case input at time k — say w_k^o — can be determined based on this information. It will be shown that the entropy of $\boldsymbol{G}$ at time k is a measure for the increase that arises in $J(\boldsymbol{w},\gamma)$ whenever $w_k \neq w_k^o$.

Recall that $\boldsymbol{M}$ is the spectral factor of the positive-definite operator $\boldsymbol{I} - \gamma^{-2}\boldsymbol{G}^*\boldsymbol{G}$. We can decompose this spectral factor as

$$\boldsymbol{M} = \boldsymbol{M}_m + \boldsymbol{M}_{sc}$$

where $\boldsymbol{M}_m$ is memoryless and $\boldsymbol{M}_{sc}$ is strictly causal.

We wish to consider the effect on the cost function $J(\boldsymbol{w},\gamma)$ that arises whenever the worst-case input is not being applied. Before doing so, we analyze the effect of general inputs on this cost function. Specifically we consider two different problems. In one, the input $\boldsymbol{w}$ is specified up to time k, and the values of w_i for $i > k$ are used to minimize $J(\boldsymbol{w},\gamma)$. In the second problem, the input is specified only up to time $k-1$, resulting in one extra degree of freedom to the optimization — the choice of w_k. We show that the gain in $J(\boldsymbol{w},\gamma)$ from this extra degree of freedom can be expressed in terms of the kth element of the memoryless part of the spectral factor.

In the following lemma, recall that $\boldsymbol{P}_k$ is the projection operator introduced in Section 2.1.

Lemma 5.1 *Suppose that $\boldsymbol{G} \in \mathcal{C}$ with $\|\boldsymbol{G}\| < \gamma$. Moreover, assume that $P_k\bar{\boldsymbol{w}}$ is given. Then, for $\boldsymbol{z} = \boldsymbol{G}\boldsymbol{w}$ we have*

$$\inf_{P_k^\perp \boldsymbol{w}} \left\{ J(\boldsymbol{w},\gamma) \middle| P_k\boldsymbol{w} = P_k\bar{\boldsymbol{w}} \right\} = \inf_{P_{k-1}^\perp \boldsymbol{w}} \left\{ J(\boldsymbol{w},\gamma) \middle| P_{k-1}\boldsymbol{w} = P_{k-1}\bar{\boldsymbol{w}} \right\} + v_k^T M_{k,k}^T M_{k,k} v_k \tag{5.1}$$

where v_k is the kth element of $\boldsymbol{P}_k(\boldsymbol{I} + \boldsymbol{M}_m^{-1}\boldsymbol{M}_{sc})\bar{\boldsymbol{w}}$.

Proof: By Lemma 2.3, $\boldsymbol{M}_m$ is invertible since $\boldsymbol{M}$ is. Hence we can write

$$\begin{aligned}\|\boldsymbol{w}\|_2^2 - \gamma^{-2}\|\boldsymbol{z}\|_2^2 &= \boldsymbol{w}^*(\boldsymbol{I} - \gamma^{-2}\boldsymbol{G}^*\boldsymbol{G})\boldsymbol{w} \\ &= \boldsymbol{w}^*\boldsymbol{M}^*\boldsymbol{M}\boldsymbol{w} \\ &= \boldsymbol{w}^*(\boldsymbol{M}_m + \boldsymbol{M}_{sc})^*(\boldsymbol{M}_m + \boldsymbol{M}_{sc})\boldsymbol{w} \\ &= \boldsymbol{w}^*(\boldsymbol{I} + \boldsymbol{M}_m^{-1}\boldsymbol{M}_{sc})^*\boldsymbol{M}_m^*\boldsymbol{M}_m(\boldsymbol{I} + \boldsymbol{M}_m^{-1}\boldsymbol{M}_{sc})\boldsymbol{w}\end{aligned}$$

By defining $\boldsymbol{v} := (\boldsymbol{I} + \boldsymbol{M}_m^{-1}\boldsymbol{M}_{sc})\boldsymbol{w}$, we can write this as

$$\begin{aligned}\|\boldsymbol{w}\|_2^2 - \gamma^{-2}\|\boldsymbol{z}\|_2^2 &= \boldsymbol{v}^*\boldsymbol{M}_m^*\boldsymbol{M}_m\boldsymbol{v} \\ &= \sum_{i=-\infty}^{\infty} v_i^T M_{i,i}^T M_{i,i} v_i\end{aligned}$$

Since the operator mapping $\boldsymbol{w}$ to $\boldsymbol{v}$ is causal, once $\boldsymbol{P}_{k-1}\boldsymbol{w}$ has been specified, $\boldsymbol{P}_{k-1}\boldsymbol{v}$ is completely determined. Moreover, it is immediate that the optimal (worst-case) $\boldsymbol{P}_{k-1}^{\perp}\boldsymbol{w}$ of the right-hand side of (5.1) occurs when $\boldsymbol{P}_{k-1}^{\perp}\boldsymbol{v} = \boldsymbol{0}$, which implies that

$$\boldsymbol{P}_{k-1}^{\perp}\boldsymbol{w} = -\boldsymbol{P}_{k-1}^{\perp}\boldsymbol{M}_m^{-1}\boldsymbol{M}_{sc}\boldsymbol{w} \tag{5.2}$$

Note that (5.2) has a unique solution since $\boldsymbol{M}_m^{-1}\boldsymbol{M}_{sc}$ is strictly causal. Now suppose that, at time k, w_k is not the worst-case disturbance, say $w_k = \bar{w}_k$. It follows that the optimal solution of the left-hand side of (5.1) satisfies

$$\boldsymbol{P}_k^{\perp}\boldsymbol{w} = -\boldsymbol{P}_k^{\perp}\boldsymbol{M}_m^{-1}\boldsymbol{M}_{sc}\boldsymbol{w}$$

again from the fact that $\boldsymbol{P}_k^{\perp}\boldsymbol{v} = \boldsymbol{0}$. The difference in value of the two optimization problems is obviously given by $v_k^T M_{k,k}^T M_{k,k} v_k$. ■

This result shows that by choosing an arbitrary w_k (not equal to the worst-case input), the increase of $J(\boldsymbol{w}, \gamma)$ can be expressed in terms of $M_{k,k}^T M_{k,k}$. By setting $\boldsymbol{P}_{k-1}\bar{\boldsymbol{w}} = \boldsymbol{0}$ in this result, we can relate the value of the entropy at time k to this increase.

Corollary 5.2 *Suppose that $\boldsymbol{G} \in \mathcal{C}$ with $\|\boldsymbol{G}\| < \gamma$ and that $\boldsymbol{z} = \boldsymbol{G}\boldsymbol{w}$. Whenever the worst-case input is applied up to time $k-1$, i.e. $\boldsymbol{P}_{k-1}\boldsymbol{w} = \boldsymbol{0}$,* *we have that*

$$\begin{aligned}
&\inf_{P_k^{\perp} w} \left\{ J(w,\gamma) \middle| P_{k-1} w = \mathbf{0} \ , \ w_k = \bar{w}_k \right\} \\
&= \inf_{P_{k-1}^{\perp} w} \left\{ J(w,\gamma) \middle| P_{k-1} w = \mathbf{0} \right\} + \bar{w}_k^T M_{k,k}^T M_{k,k} \bar{w}_k \\
&\geq \inf_{P_{k-1}^{\perp} w} \left\{ J(w,\gamma) \middle| P_{k-1} w = \mathbf{0} \right\} + \det(M_{k,k}^T M_{k,k}) |\bar{w}_k|^2 \\
&= \det(M_{k,k}^T M_{k,k}) |\bar{w}_k|^2
\end{aligned}$$

Proof: Since $P_{k-1} w = \mathbf{0}$, it follows that $P_{k-1} v = \mathbf{0}$. Moreover, we have that $v_k = \bar{w}_k$, and the first equality follows directly from Lemma 5.1. The inequality is also straightforward since $0 < M_{k,k}^T M_{k,k} \leq I$, which follows from Lemma 4.5. ■

From this result it is straightforward to obtain a direct relationship between $J(w, \gamma)$ and the entropy.

Corollary 5.3 *Suppose that $G \in \mathcal{C}$ with $\|G\| < \gamma$. Then, given the input $P_{k-1} w = \mathbf{0}$, we have*

$$\begin{aligned}
1 &\geq \frac{1}{|\bar{w}_k|^2} \inf_{P_k^{\perp} w} \left\{ J(w,\gamma) \middle| P_{k-1} w = \mathbf{0} \ , \ w_k = \bar{w}_k \right\} \\
&\geq \exp\left(-\gamma^{-2} E(G,\gamma)_k\right)
\end{aligned}$$

Proof: The first inequality follows by taking $P_k^{\perp} w = \mathbf{0}$, and the second inequality follows by rewriting the result in Corollary 5.2. ■

This corollary indicates the relationship between the entropy and $\mathcal{H}_\infty$ control, and it partly explains why a minimum entropy controller should be preferred over other $\mathcal{H}_\infty$ sub-optimal controllers. Namely, all $\mathcal{H}_\infty$ controllers which ensure that the norm bound $\|G\| < \gamma$ is satisfied, guarantee that the cost function $J(w, \gamma) > 0$ for all $\mathbf{0} \neq w \in \ell_2$. This value equals 0 if the input is the worst-case disturbance $w = \mathbf{0}$. The minimum entropy controller assures that, for the closed-loop system, $J(w, \gamma)$ increases with a guaranteed amount in the case where w is not the worst-case input, which is higher than the increase that other controllers guarantee.

Remark 5.4 Since $0 < M_{k,k}^T M_{k,k} \leq I$, the formula for the entropy, which is related to $\det(M_{k,k}^T M_{k,k})$, is a good measure for this guaranteed increase of $J(w, \gamma)$. Minimizing the smallest eigenvalue of $M_{k,k}^T M_{k,k}$ would be preferable, but this leads to a much more difficult optimization problem.

The same result can be given for the entropy of an anti-causal system $\boldsymbol{H}$. The proof is similar and will therefore be omitted.

Lemma 5.5 *Suppose that $\boldsymbol{H}^* \in \mathcal{C}$ with $\|\boldsymbol{H}\| < \gamma$. Then, given $P_k^{\perp}\boldsymbol{w} = \boldsymbol{0}$ (i.e. $\boldsymbol{P}_k^{\perp}\boldsymbol{w}$ is the worst-case input), we have for $\boldsymbol{z} = \boldsymbol{H}\boldsymbol{w}$*

$$\begin{aligned}
&\inf_{P_{k-1}\boldsymbol{w}} \left\{ \|\boldsymbol{w}\|_2^2 - \gamma^{-2}\|\boldsymbol{z}\|_2^2 \,\middle|\, P_k^{\perp}\boldsymbol{w} = \boldsymbol{0} \ , \ w_k = \bar{w}_k \right\} \\
&= \inf_{P_k\boldsymbol{w}} \left\{ \|\boldsymbol{w}\|_2^2 - \gamma^{-2}\|\boldsymbol{z}\|_2^2 \,\middle|\, P_k^{\perp}\boldsymbol{w} = \boldsymbol{0} \right\} + \bar{w}_k^T N_{k,k} N_{k,k}^T \bar{w}_k \\
&\geq \inf_{P_k\boldsymbol{w}} \left\{ \|\boldsymbol{w}\|_2^2 - \gamma^{-2}\|\boldsymbol{z}\|_2^2 \,\middle|\, P_k^{\perp}\boldsymbol{w} = \boldsymbol{0} \right\} + \det(N_{k,k}N_{k,k}^T)|\bar{w}_k|^2
\end{aligned}$$

which equals $\det(N_{k,k}N_{k,k}^T)|\bar{w}_k|^2$. □

In the next section we discuss the relationship between the entropy and the $\mathcal{H}_2$ norm.

5.2 Relationship with $\mathcal{H}_2$ control

An important property of the entropy for time-invariant systems is that it is an upper bound for the squared $\mathcal{H}_2$ norm. In order to consider this property for time-varying systems care must first be taken in choosing an appropriate quadratic cost. Consider the system with input-output operator $\boldsymbol{G} \in \mathcal{C}$. The ℓ_2 semi-norm of $\boldsymbol{G}$ (for single-infinite sequences) is the number [56]

$$\|\boldsymbol{G}\|_2^2 = \limsup_{l\to\infty} \frac{1}{l+1} \sum_{i=0}^{l} \sum_{j=0}^{i} \operatorname{tr}\left[G_{i,j}^T G_{i,j}\right] \tag{5.3}$$

Because of the non-compactness of $\boldsymbol{G}$, the normalization factor $1/(l+1)$ is introduced so as to ensure that the value in (5.3) remains bounded. For operators with a uniformly exponentially stable state-space representation (2.1), it is shown in [56] that

$$\|\boldsymbol{G}\|_2^2 = \limsup_{l\to\infty} \frac{1}{l+1} \sum_{k=0}^{l} \operatorname{tr}\left[D_k^T D_k + B_k^T L_{k+1} B_k\right]$$

where $\mathcal{M} \ni \boldsymbol{L} = \operatorname{diag}\{L_k\}$ is the positive semi-definite solution to the operator Stein equation

$$\boldsymbol{L} = \boldsymbol{A}^*\boldsymbol{Z}\boldsymbol{L}\boldsymbol{Z}^*\boldsymbol{A} + \boldsymbol{C}^*\boldsymbol{C} \tag{5.4}$$

which exists since $\boldsymbol{A}$ is UES (see Lemma 2.8).

Because we are considering entropy to be an operator, however, we must find a representation of the quadratic cost which is not a real number, but an operator instead. An alternative approach to dealing with the non-compactness of $\boldsymbol{G}$ was considered by Feintuch *et al.* [36]. Instead of (5.3), the authors consider the memoryless part of $\boldsymbol{G}^*\boldsymbol{G}$. Since this alternative cost criterion is now an operator rather than a real number, the minimization must be carried out with respect to the partial ordering of positive semi-definite operators. For our approach, we introduce a slight modification. Rather than taking $\operatorname{diag}\{(\boldsymbol{G}^*\boldsymbol{G})_{k,k}\}$ as the cost criterion, we take as our quadratic cost

$$\boldsymbol{Q}(\boldsymbol{G}) := \operatorname{diag}\Big\{\operatorname{tr}\big[(\boldsymbol{G}^*\boldsymbol{G})_{k,k}\big]\Big\} \in \mathcal{M} \tag{5.5}$$

This extra step allows us to compare the entropy (4.3) to the cost criterion (5.5). The interpretation of this quadratic cost is straightforward, as indicated by the following lemma. We say that a signal $\boldsymbol{w}$ is (zero mean) white noise at time k if $\mathcal{E}w_k = 0$, $\mathcal{E}w_k w_k^T = I$, and $w_i = 0$ for $i \neq k$.

Lemma 5.6 *Let $\boldsymbol{w}$ be white noise at time k, and zero elsewhere. Then*

$$\boldsymbol{Q}(\boldsymbol{G})_k = \mathcal{E}\,\|\boldsymbol{z}\|_2^2$$

Proof: Since $\boldsymbol{z} = \boldsymbol{G}\boldsymbol{w}$ we get

$$\begin{aligned}\mathcal{E}\,\|\boldsymbol{z}\|_2^2 &= \mathcal{E}\sum_{i=-\infty}^{\infty} w_i^T(\boldsymbol{G}^*\boldsymbol{G}\boldsymbol{w})_i \\ &= \mathcal{E}w_k^T(\boldsymbol{G}^*\boldsymbol{G}\boldsymbol{w})_k \\ &= \mathcal{E}w_k^T\sum_{j=-\infty}^{\infty}(\boldsymbol{G}^*\boldsymbol{G})_{k,j}w_j\end{aligned}$$

Since $w_j = 0$ for all $j \neq k$, this yields

$$\begin{aligned}\mathcal{E}\,\|\boldsymbol{z}\|_2^2 &= \mathcal{E}w_k^T(\boldsymbol{G}^*\boldsymbol{G})_{k,k}w_k \\ &= \mathcal{E}\operatorname{tr}\Big[(\boldsymbol{G}^*\boldsymbol{G})_{k,k}w_k w_k^T\Big] \\ &= \operatorname{tr}\big[(\boldsymbol{G}^*\boldsymbol{G})_{k,k}\big]\end{aligned}$$

which completes the proof. ∎

This quadratic cost is equivalent to the squared $\mathcal{H}_2$ cost (1.1) if the system is LTI. For systems which admit a state-space realization, this quadratic cost can be expressed in terms of the solution to a Lyapunov equation.

Lemma 5.7 *Suppose that $\boldsymbol{G}$ is given by the state-space realization (2.1), where $\boldsymbol{A}$ is UES. Then*

$$\boldsymbol{Q}(\boldsymbol{G}) = \operatorname{diag}\Big\{\operatorname{tr}\big[D_k^T D_k + B_k^T L_{k+1} B_k\big]\Big\} \tag{5.6}$$

where $\boldsymbol{L}$ is the solution to (5.4).

Proof: Using the equation (5.4) for $\boldsymbol{L}$ it is an easy calculation to write

$$\begin{aligned}\boldsymbol{G}^*\boldsymbol{G} &= \big(\boldsymbol{D} + \boldsymbol{C}(\boldsymbol{I} - \boldsymbol{Z}^*\boldsymbol{A})^{-1}\boldsymbol{Z}^*\boldsymbol{B}\big)^* \big(\boldsymbol{D} + \boldsymbol{C}(\boldsymbol{I} - \boldsymbol{Z}^*\boldsymbol{A})^{-1}\boldsymbol{Z}^*\boldsymbol{B}\big)\\ &= \boldsymbol{D}^*\boldsymbol{D} + \boldsymbol{B}^*\boldsymbol{Z}\boldsymbol{L}\boldsymbol{Z}^*\boldsymbol{B} + (\boldsymbol{B}^*\boldsymbol{Z}\boldsymbol{L}\boldsymbol{Z}^*\boldsymbol{A} + \boldsymbol{D}^*\boldsymbol{C})(\boldsymbol{I} - \boldsymbol{Z}^*\boldsymbol{A})^{-1}\boldsymbol{Z}^*\boldsymbol{B}\\ &\quad + \boldsymbol{B}^*\boldsymbol{Z}(\boldsymbol{I} - \boldsymbol{A}^*\boldsymbol{Z})^{-1}(\boldsymbol{A}^*\boldsymbol{Z}\boldsymbol{L}\boldsymbol{Z}^*\boldsymbol{B} + \boldsymbol{C}^*\boldsymbol{D})\end{aligned}$$

The third term is strictly causal, and the fourth term is strictly anti-causal. The first two terms are memoryless, and the result follows immediately. ■

For operators admitting a state-space realization, the expression for the entropy derived in Lemma 4.10 can be compared to (5.6) to relate $\boldsymbol{Q}(\boldsymbol{G})$ to $\boldsymbol{E}(\boldsymbol{G},\gamma)$. Here we will prove directly, using the definitions of these operators, that the entropy is an upper bound for the quadratic cost.

Lemma 5.8 *Suppose that $\boldsymbol{G} \in \mathcal{C}$ with $\|\boldsymbol{G}\| < \gamma$. Then $\boldsymbol{E}(\boldsymbol{G},\gamma) \geq \boldsymbol{Q}(\boldsymbol{G})$.*

Proof: From (4.7) we see that the entropy can be evaluated as

$$\boldsymbol{E}(\boldsymbol{G},\gamma)_k = \sum_{j=1}^{\infty} \frac{\gamma^{-2(j-1)}}{j} \operatorname{tr}\Big[\big(\gamma^2 I - \gamma^2 M_{k,k}^T M_{k,k}\big)^j\Big]$$

Since all the terms are positive, we get

$$\begin{aligned}\boldsymbol{E}(\boldsymbol{G},\gamma)_k &\geq \operatorname{tr}\big[\gamma^2 I - \gamma^2 M_{k,k}^T M_{k,k}\big]\\ &\geq \operatorname{tr}\Big[\gamma^2 I - \gamma^2 M_{k,k}^T M_{k,k} - \gamma^2 \sum_{i=k+1}^{\infty} M_{i,k}^T M_{i,k}\Big] \qquad (5.7)\\ &= \operatorname{tr}\left[\sum_{i=k}^{\infty} G_{i,k}^T G_{i,k}\right]\\ &= \operatorname{tr}\big[(\boldsymbol{G}^*\boldsymbol{G})_{k,k}\big]\end{aligned}$$

which equals the quadratic cost at time k. ■

It follows that the quadratic cost operator is bounded above by the entropy, as in the LTI case. Hence, minimizing the entropy automatically assures a lower upper bound for the quadratic cost. In fact, from (5.7) we can state a stronger result.

Corollary 5.9 *Suppose that $\boldsymbol{G} \in \mathcal{C}$ with $\|\boldsymbol{G}\| < \gamma$. Then*

$$\boldsymbol{Q}(\boldsymbol{G}) \leq \operatorname{diag}\left\{\gamma^2 \operatorname{tr}\left[I - M_{k,k}^T M_{k,k}\right]\right\} \qquad \square$$

The connection with the entropy is now straightforward. Minimizing this entropy at time k is equivalent to maximizing the product of the eigenvalues of $M_{k,k}^T M_{k,k}$. Since these are all in the interval $(0,1]$ (Lemma 4.5), maximizing their product will cause the sum of these eigenvalues to be higher in general. Hence the upper bound for the quadratic cost at time k, given in Corollary 5.9, will be lower. This accounts for the observation in [9] that in certain cases minimum entropy controllers result in a lower $\mathcal{H}_2$ cost than mixed $\mathcal{H}_2/\mathcal{H}_\infty$ controllers do.

We now show that, in the limit, the entropy operator will converge to the quadratic cost operator.

Lemma 5.10 *Suppose that $\boldsymbol{G} \in \mathcal{C}$. Then*

$$\lim_{\gamma \to \infty} \boldsymbol{E}(\boldsymbol{G}, \gamma) = \boldsymbol{Q}(\boldsymbol{G})$$

Proof: The following expression for the entropy was obtained in (4.7).

$$\boldsymbol{E}(\boldsymbol{G}, \gamma)_k = \sum_{j=1}^{\infty} \frac{\gamma^{-2(j-1)}}{j} \operatorname{tr}\left[(\gamma^2 I - \gamma^2 M_{k,k}^T M_{k,k})^j\right]$$

In the proof of Lemma 5.8 we showed that there are two inequalities in the comparison between this expression and the quadratic cost. In the proof of Lemma 4.7 it is shown that both $\gamma^{-2(j-1)}$ and the operator $(\gamma^2 \boldsymbol{I} - \gamma^2 \boldsymbol{M}^* \boldsymbol{M})^j$ decrease as γ increases (for all $j \geq 1$). Since both are bounded below by zero, it follows that all terms converge. For $j > 1$ we have that $\gamma^{-2(j-1)}$ goes to zero as γ does, hence we are left with the terms for $j = 1$, and the first inequality disappears, i.e.

$$\lim_{\gamma \to \infty} \boldsymbol{E}(\boldsymbol{G}, \gamma)_k = \lim_{\gamma \to \infty} \operatorname{tr}\left[\gamma^2 I - \gamma^2 M_{k,k}^T M_{k,k}\right]$$

Now, since this term converges as γ goes to infinity, it follows that $\boldsymbol{M}_m^* \boldsymbol{M}_m$ goes to the identity operator in norm as γ goes to infinity. Therefore, if we take an arbitrary vector $\bar{w}$, and compare the input

$$w_i = \begin{cases} 0 & \text{for } i < k \\ \bar{w} & \text{for } i = k \\ -M_{i,i}^{-1}\sum_{j=k}^{i-1} M_{i,j}w_j & \text{for } i > k \end{cases}$$

with the input $\boldsymbol{v}$ given by $v_k = \bar{w}$, and $v_i = 0$ for $i \neq k$. Applying the signal $\boldsymbol{w}$ to the spectral factorization, and using $(\boldsymbol{Mw})_i = 0$ for $i > k$, we get

$$\begin{aligned} |M_{k,k}\bar{w}|^2 &= \|\boldsymbol{Mw}\|_2^2 \\ &= \|\boldsymbol{w}\|_2^2 - \gamma^{-2}\|\boldsymbol{Gw}\|_2^2 \\ &= |\bar{w}|^2 + \sum_{i=k+1}^{\infty} \|w_i\|^2 - \gamma^{-2}\|\boldsymbol{Gw}\|_2^2 \end{aligned}$$

Since $M_{k,k}^T M_{k,k}$ goes to the identity as γ goes to infinity, it follows that

$$\sum_{i=k+1}^{\infty} |w_i|^2 \to 0 \quad (\gamma \to \infty) \tag{5.8}$$

Now we are ready to evaluate

$$\begin{aligned} \gamma^2\bar{w}^T \sum_{i=k+1}^{\infty} M_{i,k}^T M_{i,k}\bar{w} &= \gamma^2\bar{w}^T \sum_{i=k}^{\infty} M_{i,k}^T M_{i,k}\bar{w} - \gamma^2\bar{w}^T M_{k,k}^T M_{k,k}\bar{w} \\ &= \gamma^2\|\boldsymbol{Mv}\|_2^2 - \gamma^2\|\boldsymbol{Mw}\|_2^2 \\ &= \gamma^2\|\boldsymbol{v}\|_2^2 - \|\boldsymbol{Gv}\|_2^2 - \gamma^2\|\boldsymbol{w}\|_2^2 + \|\boldsymbol{Gw}\|_2^2 \\ &= -\gamma^2 \sum_{i=k+1}^{\infty} |w_i|^2 - \|\boldsymbol{Gv}\|_2^2 + \|\boldsymbol{Gw}\|_2^2 \end{aligned}$$

This can be bounded above by omitting the first term, resulting in

$$\begin{aligned} \gamma^2\bar{w}^T \sum_{i=k+1}^{\infty} M_{i,k}^T M_{i,k}\bar{w} &\leq -\|\boldsymbol{Gv}\|_2^2 + \|\boldsymbol{Gw}\|_2^2 \\ &= -\|\boldsymbol{Gv}\|_2^2 + \|\boldsymbol{Gv} + \boldsymbol{G}(\boldsymbol{w} - \boldsymbol{v})\|_2^2 \\ &\leq \|\boldsymbol{G}(\boldsymbol{w} - \boldsymbol{v})\|_2^2 + 2\|\boldsymbol{Gv}\|_2\|\boldsymbol{G}(\boldsymbol{w} - \boldsymbol{v})\|_2 \end{aligned} \tag{5.9}$$

From (5.8) we see that $\|\boldsymbol{w} - \boldsymbol{v}\|$ goes to zero as γ goes to infinity, and since the operator $\boldsymbol{G}$ is bounded, it follows that (5.9) goes to zero as γ goes to infinity. Since $\bar{w}$ is arbitrary we conclude that

$$\lim_{\gamma\to\infty} \gamma^2 \sum_{i=k+1}^{\infty} M_{i,k}^T M_{i,k} = 0$$

An evaluation of the entropy shows that

$$\begin{aligned}
\lim_{\gamma\to\infty} \boldsymbol{E}(\boldsymbol{G},\gamma)_k &= \lim_{\gamma\to\infty} \operatorname{tr}\left[\gamma^2 I - \gamma^2 M_{k,k}^T M_{k,k}\right] \\
&= \lim_{\gamma\to\infty} \operatorname{tr}\left[\gamma^2 I - \gamma^2 M_{k,k}^T M_{k,k} - \gamma^2 \sum_{i=k+1}^{\infty} M_{i,k}^T M_{i,k}\right] \\
&= \operatorname{tr}\left[(\boldsymbol{G}^*\boldsymbol{G})_{k,k}\right]
\end{aligned}$$

which is the quadratic cost at time k. ∎

Remark 5.11 For operators represented by a state-space realization, these results can be found in a straightforward manner. The entropy of $\boldsymbol{G}$ can be evaluated as in Lemma 4.10, as

$$\begin{aligned}
\boldsymbol{E}(\boldsymbol{G},\gamma)_k &= -\gamma^2 \ln\det\left(I - \gamma^{-2} D_k^T D_k - \gamma^{-2} B_k^T X_{k+1} B_k\right) \\
&= \sum_{j=1}^{\infty} \frac{\gamma^{-2(j-1)}}{j} \operatorname{tr}\left[D_k^T D_k + B_k^T X_{k+1} B_k\right] \qquad (5.10)
\end{aligned}$$

As is done in Claim 8 in Appendix A we can show that, given any "initial condition" $x_k = \bar{x}_k$ and $\boldsymbol{P}_{k-1}\boldsymbol{w} = \boldsymbol{0}$

$$\sup_{\boldsymbol{P}_{k-1}^{\perp}\boldsymbol{w}} \left\{ \|\boldsymbol{z}\|_2^2 - \gamma^2 \|\boldsymbol{w}\|_2^2 \right\} = \bar{x}_k^T X_k \bar{x}_k$$

Obviously, the left-hand side is a decreasing function of γ. Since $\bar{x}_k$ is arbitrary, it follows that X_k is a decreasing function of γ. Hence all the terms in (5.10) are decreasing with respect to γ, which shows that the entropy is decreasing with respect to γ, as we showed in Lemma 4.7.

Moreover, it is easily seen that the solution $\boldsymbol{X}$ of (4.9) converges to the solution $\boldsymbol{L}$ of (5.4) in norm if γ tends to infinity. Hence evaluating (5.10) gives

$$\begin{aligned}
\lim_{\gamma\to\infty} \boldsymbol{E}(\boldsymbol{G},\gamma)_k &= \lim_{\gamma\to\infty} \operatorname{tr}\left[D_k^T D_k + B_k^T X_{k+1} B_k\right] \\
&= \operatorname{tr}\left[D_k^T D_k + B_k^T L_{k+1} B_k\right] \\
&= \boldsymbol{Q}(\boldsymbol{G})_k
\end{aligned}$$

in accordance with Lemma 5.10.

In the next section we will discuss the average of several cost functions.

5.3 Average cost functions

Analogous to the quadratic cost operator defined for a causal system, we can define the quadratic cost at time k for the anti-causal operator $\boldsymbol{G}^*$.

$$\boldsymbol{Q}_a(\boldsymbol{G}^*)_k := \operatorname{tr}\left[(\boldsymbol{G}\boldsymbol{G}^*)_{k,k}\right]$$

An interpretation of this operator is straightforward. Suppose that the $\boldsymbol{w}$ is white noise at time k. Then, with $\boldsymbol{z} = \boldsymbol{G}^*\boldsymbol{w}$

$$\boldsymbol{Q}_a(\boldsymbol{G}^*)_k = \mathcal{E}\|\boldsymbol{z}\|_2^2 \leq \boldsymbol{E}_a(\boldsymbol{G}^*, \gamma)$$

where the connection with the entropy of $\boldsymbol{G}^*$ follows in the same way as for causal operators. Also it can be shown that the entropy of $\boldsymbol{G}^*$ converges to the quadratic cost associated with $\boldsymbol{G}^*$ as γ goes to infinity.

In the time-invariant case the quadratic cost of a causal operator and the quadratic cost of its anti-causal adjoint are equal. We will show that this is not the case for time-varying systems.

5.3.1 Average $\mathcal{H}_2$ cost

First we look at the connections between the average quadratic cost for a causal operator and its anti-causal adjoint. The following example illustrates the fact that the two are not equivalent.

Example 5.12 Let $\boldsymbol{G}$ be given by

$$\boldsymbol{G} = \begin{bmatrix} \boldsymbol{0} & \boldsymbol{0} \\ \boldsymbol{0} & \boldsymbol{U} \end{bmatrix}$$

where the operator $\boldsymbol{U} = \{U_{i,j}\}$, $i \geq 0$, $j \geq 0$ is given by

$$U_{i,j} = \begin{cases} \frac{1}{2} & \text{for } i = 2j \\ 0 & \text{elsewhere} \end{cases}$$

This is a bounded operator mapping ℓ_2 to ℓ_2. Simple algebra leads to $\boldsymbol{U}^*\boldsymbol{U} = \frac{1}{4}\boldsymbol{I}$, while $(\boldsymbol{U}\boldsymbol{U}^*)_{j,j} = \frac{1}{4}$ if j is even, zero elsewhere. We find

$$\limsup_{k\to\infty} \frac{1}{2k+1} \sum_{j=-k}^{k} \boldsymbol{Q}(\boldsymbol{G})_j = \lim_{k\to\infty} \frac{1}{2k+1} \sum_{j=-k}^{k} \operatorname{tr}\left[(\boldsymbol{G}^*\boldsymbol{G})_{j,j}\right] = \frac{1}{8}$$

$$\limsup_{k\to\infty} \frac{1}{2k+1} \sum_{j=-k}^{k} \boldsymbol{Q}_a(\boldsymbol{G}^*)_j = \lim_{k\to\infty} \frac{1}{2k+1} \sum_{j=-k}^{k} \operatorname{tr}\left[(\boldsymbol{G}\boldsymbol{G}^*)_{j,j}\right] = \frac{1}{16}$$

which are not equal to each other. □

For systems which admit a state-space realization (with bounded state-space matrices), we know that $\|G_{k,l}\| < c\beta^{k-l}$ for $k \geq l$ if $\boldsymbol{G}$ is stable (see Section 2.2). Denote by $\boldsymbol{w}_{[-k,k]}$ an input sequence consisting of white noise w_i for $i \in \{-k, \cdots, k\}$ and zero elsewhere, that is, if

$$v := \begin{bmatrix} w_{-k} \\ \vdots \\ w_k \end{bmatrix} \tag{5.11}$$

then $\mathcal{E}v = 0$ and $\mathcal{E}vv^T = I_{(2k+1)m}$, where m is the dimension of the input $\boldsymbol{w}$. Now, decompose the operator $\boldsymbol{G}$ according to this partition of time:

$$\boldsymbol{G} = \begin{bmatrix} \boldsymbol{G}_{[-,-]} & \boldsymbol{0} & \boldsymbol{0} \\ \boldsymbol{G}_{[k,-]} & G_{[k,k]} & \boldsymbol{0} \\ \boldsymbol{G}_{[+,-]} & \boldsymbol{G}_{[+,k]} & \boldsymbol{G}_{[+,+]} \end{bmatrix} \tag{5.12}$$

Here $G_{[k,k]}$ is a $(2k+1) \times (2k+1)$ block matrix, given by $G_{[k,k]} = \{G_{i,j}\}$ for $-k \leq i, j \leq k$. The other elements of (5.12) are all infinite-dimensional operators.

We now provide two preliminary results. The first will be used throughout the sequel. The second equates the average quadratic cost of $\boldsymbol{G}$ and $\boldsymbol{G}^*$ to each other.

Lemma 5.13 *Suppose that $\boldsymbol{G} \in \mathcal{C}$ admits a stable state-space realization. Then*

$$\lim_{k \to \infty} \frac{1}{2k+1} \sum_{j=-k}^{k} \operatorname{tr}\left[\left(\boldsymbol{G}^*_{[+,k]} \boldsymbol{G}_{[+,k]} \right)_{j,j} \right] = 0$$

Proof: Each element of $\boldsymbol{G}$ satisfies $\|G_{k,l}\| \leq c\beta^{k-l}$ for some constants c and $\beta \in [0,1)$, which depend on the matrices in the realization. Now, for $-k \leq j \leq k$ we can write

$$\left(\boldsymbol{G}^*_{[+,k]} \boldsymbol{G}_{[+,k]} \right)_{j,j} = \sum_{i=k+1}^{\infty} G^T_{i,j} G_{i,j}$$

Using the norm bound obtained above on the elements of $G_{i,j}$, we obtain

$$\operatorname{tr}\left[\left(\boldsymbol{G}^*_{[+,k]} \boldsymbol{G}_{[+,k]} \right)_{j,j} \right] \leq \operatorname{tr}\left[\sum_{i=k+1}^{\infty} c^2 \beta^{2(i-j)} I \right] = \frac{mc^2 \beta^{2(k+1-j)}}{1-\beta^2}$$

Therefore

$$0 \leq \lim_{k\to\infty} \frac{1}{2k+1} \sum_{j=-k}^{k} \operatorname{tr}\left[\left(\boldsymbol{G}^*_{[+,k]}\boldsymbol{G}_{[+,k]}\right)_{j,j}\right]$$

$$\leq \lim_{k\to\infty} \frac{mc^2\beta^2(1-\beta^{2(2k+1)})}{(2k+1)(1-\beta^2)^2}$$

The right-hand side equals zero, and the result follows. ∎

Lemma 5.14 *Suppose that $\boldsymbol{G} \in \mathcal{C}$ admits a stable state-space realization. Then*

$$\limsup_{k\to\infty} \frac{1}{2k+1} \sum_{j=-k}^{k} \boldsymbol{Q}(\boldsymbol{G})_j = \limsup_{k\to\infty} \frac{1}{2k+1} \sum_{j=-k}^{k} \boldsymbol{Q}_a(\boldsymbol{G}^*)_j \tag{5.13}$$

Proof: As in the proof of Lemma 5.13 we get

$$\lim_{k\to\infty} \frac{1}{2k+1} \sum_{j=-k}^{k} \operatorname{tr}\left[\left(\boldsymbol{G}_{[k,-]}\boldsymbol{G}^*_{[k,-]}\right)_{j,j}\right] = 0 \tag{5.14}$$

With the decomposition (5.12), we see that the influence of a white noise signal on $\{-k, \cdots, k\}$ on the output is given by

$$\limsup_{k\to\infty} \frac{1}{2k+1} \sum_{j=-k}^{k} \boldsymbol{Q}(\boldsymbol{G})_j$$

$$= \limsup_{k\to\infty} \frac{1}{2k+1} \sum_{j=-k}^{k} \operatorname{tr}\left[\left(G^T_{[k,k]}G_{[k,k]}\right)_{j,j} + \left(\boldsymbol{G}^*_{[+,k]}\boldsymbol{G}_{[+,k]}\right)_{j,j}\right]$$

From Lemma 5.13, the term involving $\boldsymbol{G}_{[+,k]}$ is zero. In the remaining term, the order of the $G_{[k,k]}$ and its transpose can be switched, due to the trace operator. Moreover, from (5.14), we can add terms corresponding to $\boldsymbol{G}_{[k,-]}$ so that

$$\limsup_{k\to\infty} \frac{1}{2k+1} \sum_{j=-k}^{k} \boldsymbol{Q}(\boldsymbol{G})_j$$

$$= \limsup_{k\to\infty} \frac{1}{2k+1} \sum_{j=-k}^{k} \operatorname{tr}\left[\left(G_{[k,k]}G^T_{[k,k]}\right)_{j,j} + \left(\boldsymbol{G}_{[k,-]}\boldsymbol{G}^*_{[k,-]}\right)_{j,j}\right]$$

$$= \limsup_{k\to\infty} \frac{1}{2k+1} \sum_{j=-k}^{k} \boldsymbol{Q}_a(\boldsymbol{G}^*)_j$$

as required. ∎

Remark 5.15 In Lemma 5.14 we have used *limsup* instead of *lim* because, even for a stable state-space realization, the limit in (5.13) need not exist. For example take a realization $\boldsymbol{G} : \ell_2^1 \to \ell_2^1$ with only a bounded direct feedthrough term (i.e. $\boldsymbol{G} = \boldsymbol{D}$), for which the average of the scalar D_k elements does not converge.

5.3.2 Average entropy

For the time-varying entropy we can derive properties similar to those of the previous section. First we will show that the average entropy of $\boldsymbol{G}$ does not have to be equal to the average entropy of $\boldsymbol{G}^*$.

Example 5.16 Consider the operators given in Example 5.12. With $\gamma = 1$ we get

$$\boldsymbol{I} - \boldsymbol{G}^*\boldsymbol{G} = \begin{bmatrix} \boldsymbol{I} & \boldsymbol{0} \\ \boldsymbol{0} & \frac{3}{4}\boldsymbol{I} \end{bmatrix} \quad \text{and} \quad \boldsymbol{I} - \boldsymbol{G}\boldsymbol{G}^* = \begin{bmatrix} \boldsymbol{I} & \boldsymbol{0} \\ \boldsymbol{0} & \boldsymbol{I} - \boldsymbol{U}\boldsymbol{U}^* \end{bmatrix}$$

where $\boldsymbol{I} - \boldsymbol{U}\boldsymbol{U}^*$ is a memoryless operator with $(\boldsymbol{I} - \boldsymbol{U}\boldsymbol{U}^*)_{j,j} = \frac{3}{4}$ when j is even, and 1 when j is odd. Since these operators are memoryless, it straightforward to compute the entropies:

$$\boldsymbol{E}(\boldsymbol{G}, 1) = \begin{bmatrix} \boldsymbol{0} & \boldsymbol{0} \\ \boldsymbol{0} & \ln \frac{4}{3}\boldsymbol{I} \end{bmatrix} \quad \text{and} \quad \boldsymbol{E}_a(\boldsymbol{G}^*, 1) = \begin{bmatrix} \boldsymbol{0} & \boldsymbol{0} \\ \boldsymbol{0} & \boldsymbol{T} \end{bmatrix}$$

where $\boldsymbol{T}$ is a memoryless operator with $T_{j,j} = \ln \frac{4}{3}$ if j is even, and 0 if j is odd. This example shows that the averages of these entropies are not equal. An easy calculation gives

$$\limsup_{k\to\infty} \frac{1}{2k+1} \sum_{j=-k}^{k} \boldsymbol{E}(\boldsymbol{G})_j = \lim_{k\to\infty} -\frac{k+1}{2k+1} \ln \frac{3}{4} = \frac{1}{2} \ln \frac{4}{3}$$

$$\limsup_{k\to\infty} \frac{1}{2k+1} \sum_{j=-k}^{k} \boldsymbol{E}_a(\boldsymbol{G}^*)_j = \lim_{l\to\infty} -\frac{l+1}{4l+1} \ln \frac{3}{4} = \frac{1}{4} \ln \frac{4}{3}$$

where $k = 2l$. □

For operators that can be represented by a stable state-space realization however, the average entropy of a causal system does equal the entropy of its anti-causal adjoint.

Lemma 5.17 *Suppose that $\boldsymbol{G} \in \mathcal{C}$ satisfies $\|\boldsymbol{G}\| < \gamma$, and that it admits a state-space realization. Then*

$$\limsup_{k\to\infty} \frac{1}{2k+1} \sum_{j=-k}^{k} \boldsymbol{E}(\boldsymbol{G},\gamma)_j = \limsup_{k\to\infty} \frac{1}{2k+1} \sum_{j=-k}^{k} \boldsymbol{E}_a(\boldsymbol{G}^*,\gamma)_j$$

Proof: By definition of the entropy we have

$$\begin{aligned}\sum_{j=-k}^{k} \boldsymbol{E}(\boldsymbol{G},\gamma)_j &= \sum_{j=-k}^{k} -\gamma^2 \ln\det\left(M_{j,j}^T M_{j,j}\right) \\ &= -\gamma^2 \ln \prod_{j=-k}^{k} \det\left(M_{j,j}^T M_{j,j}\right) \\ &= -\gamma^2 \ln\det\left(M_{[k,k]}^T M_{[k,k]}\right)\end{aligned}$$

To evaluate this, denote

$$\begin{aligned}V &:= I - \gamma^{-2} G_{[k,k]}^T G_{[k,k]} \\ W &:= \gamma^{-2} \boldsymbol{G}_{[+,k]}^* \boldsymbol{G}_{[+,k]} + \boldsymbol{M}_{[+,k]}^* \boldsymbol{M}_{[+,k]}\end{aligned}$$

Since $\boldsymbol{I} - \boldsymbol{G}^*\boldsymbol{G} > \boldsymbol{0}$ (say $\boldsymbol{I} - \boldsymbol{G}^*\boldsymbol{G} \geq \nu I$ for some $\nu > 0$), it is immediate that

$$I \geq V \geq I - \gamma^{-2} G_{[k,k]}^T G_{[k,k]} - \gamma^{-2} \boldsymbol{G}_{[+,k]}^* \boldsymbol{G}_{[+,k]} \geq \nu I \tag{5.15}$$

Hence the matrix V is invertible. Using the spectral factorization we can write

$$\begin{aligned}M_{[k,k]}^T M_{[k,k]} &= I - \gamma^{-2} G_{[k,k]}^T G_{[k,k]} - \gamma^{-2} \boldsymbol{G}_{[+,k]}^* \boldsymbol{G}_{[+,k]} - \boldsymbol{M}_{[+,k]}^* \boldsymbol{M}_{[+,k]} \\ &= V - W\end{aligned}$$

Thus

$$\begin{aligned}\sum_{j=-k}^{k} \boldsymbol{E}(\boldsymbol{G},\gamma)_j &= -\gamma^2 \ln\det(V - W) \\ &= -\gamma^2 \ln\det(V) - \gamma^2 \ln\det(I - V^{-1/2} W V^{-1/2})\end{aligned} \tag{5.16}$$

Furthermore, since $\boldsymbol{M} \in \mathcal{C} \cap \mathcal{C}^{-1}$, it can be checked (as in Lemma 2.3) that

$$M_{[k,k]}^T M_{[k,k]} \geq \eta I$$

independent of k for some $\eta > 0$. Together with $V \leq I$ we get

$$I \geq I - V^{-1/2}WV^{-1/2} = V^{-1/2}M_{[k,k]}^T M_{[k,k]}V^{-1/2} \geq \eta I$$

Hence, both terms in (5.16) are well-defined and non-negative. Consider the second term:

$$V^{-1/2}WV^{-1/2} = I - V^{-1/2}M_{[k,k]}^T M_{[k,k]}V^{-1/2} \leq I - \eta V^{-1} \leq (1-\eta)I$$

Thus the $(2k+1)m$ eigenvalues of the matrix $V^{-1/2}WV^{-1/2}$, denoted by $\lambda_1, \cdots, \lambda_{(2k+1)m}$, are all in the interval $[0, 1-\eta]$. A standard argument shows that those eigenvalues satisfy

$$-\ln(1-\lambda_i) \leq \eta^{-1}\lambda_i \tag{5.17}$$

Since both $\boldsymbol{G}$ and $\boldsymbol{M}$ correspond to a stable state-space realization, it follows, as in the proof of Lemma 5.13, that

$$\begin{aligned}
&\lim_{k\to\infty} \frac{1}{2k+1} \operatorname{tr} W \\
&\quad = \lim_{k\to\infty} \frac{\gamma^{-2}}{2k+1} \sum_{j=-k}^{k} \operatorname{tr}\left[\left(\boldsymbol{G}_{[+,k]}^* \boldsymbol{G}_{[+,k]}\right)_{j,j} + \left(\boldsymbol{M}_{[+,k]}^* \boldsymbol{M}_{[+,k]}\right)_{j,j}\right] \\
&\quad = 0
\end{aligned}$$

Hence, using (5.17), we get

$$\begin{aligned}
&\lim_{k\to\infty} -\frac{\gamma^2}{2k+1} \ln\det\left(I - V^{-1/2}WV^{-1/2}\right) \\
&\quad = \lim_{k\to\infty} -\frac{\gamma^2}{2k+1} \ln \prod_{j=1}^{(2k+1)m} (1-\lambda_j) \\
&\quad = \lim_{k\to\infty} -\frac{\gamma^2}{2k+1} \sum_{j=1}^{(2k+1)m} \ln(1-\lambda_j) \\
&\quad \leq \eta^{-1}\gamma^2 \lim_{k\to\infty} \frac{1}{2k+1} \sum_{j=1}^{(2k+1)m} \lambda_j
\end{aligned}$$

Since the eigenvalues of $V^{-1/2}WV^{-1/2}$ are the same as the eigenvalues of $W^{1/2}V^{-1}W^{1/2}$, we obtain using (5.15)

$$\begin{aligned}
&\lim_{k\to\infty} -\frac{\gamma^2}{2k+1}\ln\det\left(I - V^{-1/2}WV^{-1/2}\right) \\
&= \eta^{-1}\gamma^2 \lim_{k\to\infty}\frac{1}{2k+1}\operatorname{tr}\left[W^{1/2}V^{-1}W^{1/2}\right] \\
&\leq \eta^{-1}\nu^{-1}\gamma^2 \lim_{k\to\infty}\frac{1}{2k+1}\operatorname{tr} W \\
&= 0
\end{aligned}$$

Similar arguments can be used for the entropy of the anti-causal operator $\boldsymbol{G}^*$, to obtain

$$\begin{aligned}
&\limsup_{k\to\infty}\frac{1}{2k+1}\sum_{j=-k}^{k}\boldsymbol{E}_a(\boldsymbol{G}^*,\gamma)_j \\
&= \limsup_{k\to\infty} -\frac{\gamma^2}{2k+1}\ln\det\left(I - \gamma^{-2}G_{[k,k]}G^T_{[k,k]}\right) \\
&= \limsup_{k\to\infty} -\frac{\gamma^2}{2k+1}\ln\det\left(I - \gamma^{-2}G^T_{[k,k]}G_{[k,k]}\right) \\
&= \limsup_{k\to\infty}\frac{1}{2k+1}\sum_{j=-k}^{k}\boldsymbol{E}(\boldsymbol{G},\gamma)_j
\end{aligned}$$

which completes the proof. ■

In the next section we consider the connection between the time-varying entropy and a time-varying analogue of the risk-sensitive problem.

5.4 Time-varying risk-sensitive control

For LTI systems, it was shown in Chapter 1 that the entropy has a nice interpretation as an $\mathcal{H}_2$ cost for which large deviations of the output are penalized.

To investigate this property for our definition of the time-varying entropy, we must first extend the concept of risk-sensitive control for time-varying systems. For a causal operator $\boldsymbol{G}$ mapping $\boldsymbol{w}$ to $\boldsymbol{z}$, define the time-varying LEQG operator as follows:

Definition 5.18 *Let w be a sequence consisting of Gaussian white noise at k, and zero elsewhere. Then*

$$\boldsymbol{R}(\boldsymbol{G},\theta) := -\frac{2}{\theta}\operatorname{diag}\left\{\ln \mathcal{E}\exp\left(-\frac{\theta}{2}\|z\|_2^2\right)\right\}$$

We will investigate this operator for negative values of θ, which is the so-called risk-averse case. To evaluate the LEQG operator, we use a standard derivation. For a signal $y = Fv$, where $v \in \mathbb{R}^m$ is a Gaussian random variable with zero expectation and unit variance, we can compute

$$\begin{aligned}
\mathcal{E}\exp\left(-\frac{\theta}{2}y^Ty\right) &= \frac{1}{\sqrt{(2\pi)^m}}\int_{\mathbb{R}^m}\exp\left(-\frac{\theta}{2}v^TF^TFv\right)\exp\left(-\frac{1}{2}v^Tv\right)dv \\
&= \frac{1}{\sqrt{(2\pi)^m}}\int_{\mathbb{R}^m}\exp\left(-\frac{1}{2}v^T\left(I+\theta F^TF\right)v\right)dv \\
&= \frac{1}{\sqrt{(2\pi)^m}}\int_{\mathbb{R}^m}\exp\left(-\frac{1}{2}r^Tr\right)\det\left|I+\theta F^TF\right|^{-1/2}dr \\
&= \det\left|I+\theta F^TF\right|^{-1/2} \qquad (5.18)
\end{aligned}$$

Using (5.18), we can evaluate the LEQG operator. If w_i is Gaussian white noise at $i = k$, and zero elsewhere, the output equals

$$z_i = \begin{cases} 0 & \text{for } i < k \\ G_{i,k}w_k & \text{for } i \geq k \end{cases}$$

Using the spectral factorization $\boldsymbol{I} - \gamma^{-2}\boldsymbol{G}^*\boldsymbol{G} = \boldsymbol{M}^*\boldsymbol{M}$, we see that

$$\frac{1}{2\gamma^2}\|\boldsymbol{z}\|_2^2 - \frac{1}{2}\|\boldsymbol{w}\|_2^2 = -\frac{1}{2}w_k^T\left(\sum_{i=k}^{\infty}M_{i,k}^TM_{i,k}\right)w_k$$

Thus we get, as in (5.18)

$$\begin{aligned}
\boldsymbol{R}(\boldsymbol{G},-\gamma^{-2})_k &= 2\gamma^2\ln\mathcal{E}\exp\left(\frac{1}{2\gamma^2}\|\boldsymbol{z}\|_2^2\right) \\
&= 2\gamma^2\ln\det\left(\sum_{i=k}^{\infty}M_{i,k}^TM_{i,k}\right)^{-1/2} \\
&= -\gamma^2\ln\det\left(\sum_{i=k}^{\infty}M_{i,k}^TM_{i,k}\right) \\
&\leq -\gamma^2\ln\det\left(M_{k,k}^TM_{k,k}\right) \\
&= E(\boldsymbol{G},\gamma)_k
\end{aligned}$$

Unlike the LTI case (Chapter 1), equality is not achieved. Nevertheless, if we take a Gaussian white noise signal as input, the *average* LEQG cost equals the *average* entropy.

Remark 5.19 The memoryless part of the spectral factor satisfies the property that

$$-\gamma^2 \ln\det\left(M_{[k,k]}^T M_{[k,k]}\right) = -\gamma^2 \sum_{i=-k}^{k} \ln\det\left(M_{i,i}^T M_{i,i}\right)$$

The quadratic cost operator satisfies a similar property; specifically

$$\mathcal{E}|z_k|^2\Big|_{\boldsymbol{w} \text{ white noise}} = \sum_{i=-\infty}^{\infty} \mathcal{E}|z_k|^2\Big|_{w_j \text{ white noise at } j=i, \text{ zero elsewhere}}$$

The LEQG cost however, does *not* satisfy this property. Therefore there is a significant difference in choice of $\boldsymbol{w}$.

Theorem 5.20 *Suppose that $\boldsymbol{G} \in \mathcal{C}$ has norm bound $\|\boldsymbol{G}\| < \gamma$, and that it admits a state-space realization. Let $\boldsymbol{w}$ be Gaussian white noise on $\{-k, \cdots, k\}$, and zero elsewhere. Then*

$$\limsup_{k\to\infty} \frac{2\gamma^2}{2k+1} \ln \mathcal{E} \exp\left(\frac{1}{2\gamma^2} \|z\|_2^2\right) = \limsup_{k\to\infty} \frac{1}{2k+1} \sum_{j=-k}^{k} \boldsymbol{E}(\boldsymbol{G}, \gamma)_j \tag{5.19}$$

Proof: Using the decomposition (5.12) we can evaluate the output as

$$\boldsymbol{z} = \boldsymbol{G}\boldsymbol{w} = \begin{bmatrix} \boldsymbol{0} \\ G_{[k,k]} \\ \boldsymbol{G}_{[+,k]} \end{bmatrix} v$$

where v is given by (5.11) (i.e. the non-zero component of $\boldsymbol{w}$). The spectral factorization gives

$$I - \gamma^{-2} G_{[k,k]}^T G_{[k,k]} - \gamma^{-2} \boldsymbol{G}_{[+,k]}^* \boldsymbol{G}_{[+,k]} = M_{[k,k]}^T M_{[k,k]} + \boldsymbol{M}_{[+,k]}^* \boldsymbol{M}_{[+,k]} \tag{5.20}$$

Hence, we can evaluate as in (5.18)

$$\begin{aligned}
&2\gamma^2 \ln \mathcal{E} \exp\Big(\frac{1}{2\gamma^2} \|z\|_2^2\Big) \\
&\quad = 2\gamma^2 \ln \det \left(M_{[k,k]}^T M_{[k,k]} + \boldsymbol{M}_{[+,k]}^* \boldsymbol{M}_{[+,k]}\right)^{-1/2} \qquad (5.21)\\
&\quad = -\gamma^2 \ln \det \left(M_{[k,k]}^T M_{[k,k]}\right) \\
&\qquad - \gamma^2 \ln \det \left(I + M_{[k,k]}^{-T} \boldsymbol{M}_{[+,k]}^* \boldsymbol{M}_{[+,k]} M_{[k,k]}^{-1}\right) \\
&\quad = \sum_{j=-k}^{k} \boldsymbol{E}(\boldsymbol{G},\gamma)_j - \gamma^2 \ln \det \left(I + M_{[k,k]}^{-T} \boldsymbol{M}_{[+,k]}^* \boldsymbol{M}_{[+,k]} M_{[k,k]}^{-1}\right)
\end{aligned}$$

The second term is obviously less than or equal to zero, hence the average entropy is an upper bound for the average LEQG cost. To show that they are equal, we have to show that the average of the second term tends to zero. Since $\boldsymbol{M} \in \mathcal{C} \cap \mathcal{C}^{-1}$, it can be shown (as in Lemma 2.3) that there exists $\eta > 0$ such that $M_{[k,k]}^T M_{[k,k]} \geq \eta I$ independent of k. Since we do not want to talk about the determinant of an infinite-dimensional operator, we factor $\boldsymbol{M}_{[+,k]}^* \boldsymbol{M}_{[+,k]} = S^T S$, where S is a finite-dimensional matrix. Using this we can evaluate

$$\begin{aligned}
&-\gamma^2 \ln \det \left(I + (M_{[k,k]}^{-T} \boldsymbol{M}_{[+,k]}^* \boldsymbol{M}_{[+,k]} M_{[k,k]}^{-1}\right) \qquad (5.22)\\
&\quad = -\gamma^2 \ln \det \left(I + S \left(M_{[k,k]}^T M_{[k,k]}\right)^{-1} S^T\right) \\
&\quad \geq -\gamma^2 \ln \det (I + \eta^{-1} S S^T) \\
&\quad = -\gamma^2 \ln \det \left(I + \eta^{-1} \boldsymbol{M}_{[+,k]}^* \boldsymbol{M}_{[+,k]}\right)
\end{aligned}$$

Denote by $\lambda_1, \cdots, \lambda_{(2k+1)m}$ the $(2k+1)m$ non-negative eigenvalues of the positive semi-definite matrix $\boldsymbol{M}_{[+,k]}^* \boldsymbol{M}_{[+,k]}$. Hence

$$-\gamma^2 \ln \det \left(I + \eta^{-1} \boldsymbol{M}_{[+,k]}^* \boldsymbol{M}_{[+,k]}\right) = -\gamma^2 \ln \prod_{j=1}^{(2k+1)m} (1 + \eta^{-1} \lambda_j)$$

Since $\ln(1+x) \leq x$ for $x \geq 0$, this expression can be bounded below by

$$
\begin{aligned}
-\gamma^2 \ln\det\left(I + \eta^{-1}\boldsymbol{M}^*_{[+,k]}\boldsymbol{M}_{[+,k]}\right) \\
&= -\gamma^2 \sum_{j=1}^{(2k+1)m} \ln(1+\eta^{-1}\lambda_j) \\
&\geq -\eta^{-1}\gamma^2 \sum_{j=1}^{(2k+1)m} \lambda_j \\
&= -\eta^{-1}\gamma^2 \sum_{j=-k}^{k} \operatorname{tr}\left[\left(\boldsymbol{M}^*_{[+,k]}\boldsymbol{M}_{[+,k]}\right)_{j,j}\right]
\end{aligned}
$$

By dividing by $2k+1$ and taking the limit as k goes to infinity, it follows from Lemma 5.13 that this term goes to zero. ■

Remark 5.21 We have considered the output z defined over all $\mathbb{Z}$. If we restrict the output to a finite horizon, say $\{-k, \cdots, k\}$, we obtain the same result. Namely, (5.20) can be written as

$$
I - \gamma^{-2} G^T_{[k,k]} G_{[k,k]} = M^T_{[k,k]} M_{[k,k]} + \boldsymbol{M}^*_{[+,k]}\boldsymbol{M}_{[+,k]} + \gamma^{-2}\boldsymbol{G}^*_{[+,k]}\boldsymbol{G}_{[+,k]}
$$

and we can use similar arguments to those of Theorem 5.20 for this expression. This will be considered in Section 5.5.

The proof of Theorem 5.20 uses the stability property of a state-space realization. The following example shows that, if the operator $\boldsymbol{G}$ can not be represented by a state-space realization, the average LEQG cost does not have to be equal to the average entropy. Note that, if we do not have a state-space realization, the limit of the distribution in the left-hand side of (5.19) might not exist. In the example we will evaluate the cost function generated by (5.21), which represents the left-hand side of (5.19).

Example 5.22 Let $\boldsymbol{G}$ be given by

$$
\boldsymbol{G} = \begin{bmatrix} \boldsymbol{0} & \boldsymbol{0} \\ \boldsymbol{0} & \sqrt{\tfrac{7}{16}\boldsymbol{I} + \tfrac{1}{4}\boldsymbol{U}} \end{bmatrix}
$$

where $\boldsymbol{U}$ was defined in Example 5.12. Note that $\boldsymbol{G}$ is a bounded operator from ℓ_2 to ℓ_2, and that $\boldsymbol{R}^*\boldsymbol{R} = \boldsymbol{I}$. Since

$$\begin{aligned}
\boldsymbol{I} - \left(\sqrt{\frac{7}{16}}\boldsymbol{I} + \frac{1}{4}\boldsymbol{U}\right)^* \left(\sqrt{\frac{7}{16}}\boldsymbol{I} + \frac{1}{4}\boldsymbol{U}\right) \\
= \frac{1}{2}\boldsymbol{I} - \frac{1}{4}\sqrt{\frac{7}{16}}\boldsymbol{U} - \frac{1}{4}\sqrt{\frac{7}{16}}\boldsymbol{U}^* \\
= \left(\sqrt{\frac{7}{16}}\boldsymbol{I} - \frac{1}{4}\boldsymbol{U}\right)^* \left(\sqrt{\frac{7}{16}}\boldsymbol{I} - \frac{1}{4}\boldsymbol{U}\right)
\end{aligned}$$

a spectral factorization $\boldsymbol{I} - \boldsymbol{G}^*\boldsymbol{G} = \boldsymbol{M}^*\boldsymbol{M}$ is given by

$$\boldsymbol{M} = \begin{bmatrix} \boldsymbol{I} & \boldsymbol{0} \\ \boldsymbol{0} & \sqrt{\frac{7}{16}}\boldsymbol{I} - \frac{1}{4}\boldsymbol{U} \end{bmatrix}$$

if $\gamma = 1$. All we have to show is that the average of (5.22) does not tend to zero. It is immediate from the definition of a spectral factorization that $M_{[k,k]}^T M_{[k,k]} \leq I$. Hence, using similar arguments as in the proof of Theorem 5.20, we find

$$\begin{aligned}
-\ln\det\left(I + M_{[k,k]}^{-T}\boldsymbol{M}_{[+,k]}^*\boldsymbol{M}_{[+,k]}M_{[k,k]}^{-1}\right) &\leq -\ln\det(I + SS^T) \\
&= -\ln\det\left(I + \boldsymbol{M}_{[+,k]}^*\boldsymbol{M}_{[+,k]}\right)
\end{aligned}$$

Suppose that k is odd. From the spectral factor $\boldsymbol{M}$ we can compute the $(2k+1) \times (2k+1)$ matrix

$$\boldsymbol{M}_{[+,k]}^*\boldsymbol{M}_{[+,k]} = \begin{bmatrix} 0 & 0 \\ 0 & \frac{1}{16}I \end{bmatrix}$$

where the identity matrix has dimension $\frac{k+1}{2} \times \frac{k+1}{2}$. Therefore

$$\begin{aligned}
-\ln\det\left(I + M_{[k,k]}^{-T}\boldsymbol{M}_{[+,k]}^*\boldsymbol{M}_{[+,k]}M_{[k,k]}^{-1}\right) &\leq -\ln\det\left(I + \boldsymbol{M}_{[+,k]}^*\boldsymbol{M}_{[+,k]}\right) \\
&= -\ln\left(\frac{17}{16}\right)^{(k+1)/2} \\
&= -\frac{k+1}{2}\ln\frac{17}{16}
\end{aligned}$$

By dividing by $2k+1$, and taking the limit as k goes to infinity, this goes to $-\frac{1}{4}\ln\frac{17}{16}$, which is strictly negative. □

In the next section we consider connections between the several control problems on a finite horizon.

5.5 Problems defined on a finite horizon

In the previous sections we discussed connections between the entropy, the quadratic cost, and the LEQG cost for systems defined on an infinite horizon. By considering these cost functions over a finite horizon, we can deal with control problems which are computationally tractable. Technically, they are also simpler since the issue of stability does not arise. Note that even for LTI systems, the possible connections between LEQG and entropy are unknown in this case, as the results of [40] are asymptotic results as the horizon goes to infinity.

Consider the LEQG and entropy cost operators on a finite interval, say $\{-k, \cdots, k\}$. Both the input (say $w|_{-k}^{k}$) and the output (say $z|_{-k}^{k}$) are restricted to that interval. From the decomposition (5.12) we get the relationship between those signals as

$$z|_{-k}^{k} = G_{[k,k]} w|_{-k}^{k}$$

Now, using the spectral factorization $\boldsymbol{I} - \gamma^{-2}\boldsymbol{G}^*\boldsymbol{G} = \boldsymbol{M}^*\boldsymbol{M}$, we see that (see (5.20))

$$I - \gamma^{-2} G_{[k,k]}^T G_{[k,k]} \neq M_{[k,k]}^T M_{[k,k]}$$

Instead of using the spectral factorization with respect to $\boldsymbol{G}$, compute the Cholesky factorization

$$I - \gamma^{-2} G_{[k,k]}^T G_{[k,k]} = L^T L$$

where L is a lower triangular matrix which has a lower triangular inverse. The differences between this finite horizon approach and that of the previous sections are significant. On a finite horizon, the average of each cost function for a causal system does turn out to be equal to the average of that cost function for its anti-causal adjoint. As an example we take the quadratic cost.

Example 5.23 Consider the operators given in Example 5.12. For k odd we get the $(2k+1) \times (2k+1)$ matrices

$$G_{[k,k]}^T G_{[k,k]} = \begin{bmatrix} 0 & 0 \\ 0 & S \end{bmatrix} \quad \text{and} \quad G_{[k,k]} G_{[k,k]}^T = \begin{bmatrix} 0 & 0 \\ 0 & T \end{bmatrix}$$

where S and T are $(k+1) \times (k+1)$ diagonal matrices given by

$$S_{j,j} = \begin{cases} \frac{1}{4} & \text{for } j \leq \frac{k+1}{2} \\ 0 & \text{elsewhere} \end{cases} \quad \text{and} \quad T_{j,j} = \begin{cases} \frac{1}{4} & \text{if } j \text{ is even} \\ 0 & \text{elsewhere} \end{cases}$$

Hence, we get

$$\sum_{j=-k}^{k} Q(G_{[k,k]})_j = \sum_{j=-k}^{k} Q(G_{[k,k]}^T)_j = \frac{k+1}{8}$$

which also follows immediately from $\operatorname{tr}\left[G_{[k,k]}^T G_{[k,k]}\right] = \operatorname{tr}\left[G_{[k,k]} G_{[k,k]}^T\right]$. □

This illustrates the difficulties that arise from the use of infinite-dimensional operators, difficulties that do not occur on a finite horizon. In the latter case, an evaluation of the cost functionals results in the following expression for the entropy

$$\begin{aligned}\sum_{j=-k}^{k} E(G_{[k,k]},\gamma)_j &= \sum_{j=-k}^{k} -\gamma^2 \ln\det(L_{j,j}^T L_{j,j}) \\ &= -2\gamma^2 \ln \prod_{j=-k}^{k} \det(L_{j,j}) \\ &= -2\gamma^2 \ln\det(L) \\ &= -\gamma^2 \ln\det(L^T L)\end{aligned}$$

From (5.18) we know that the LEQG cost can be written as

$$2\gamma^2 \ln \mathcal{E} \exp\Big(\frac{1}{2\gamma^2} \sum_{j=-k}^{k} |z_j|^2\Big) = -\gamma^2 \ln\det\left(L^T L\right)$$

And the quadratic cost can be evaluated as

$$\begin{aligned}\sum_{j=-k}^{k} Q(G_{[k,k]})_j &= \sum_{j=-k}^{k} \operatorname{tr}\left[\left(G_{[k,k]}^T G_{[k,k]}\right)_{j,j}\right] \\ &= \gamma^2 \sum_{j=-k}^{k} \operatorname{tr}\left[I - \left(L^T L\right)_{j,j}\right] \\ &= \gamma^2 \operatorname{tr}\left[I - L^T L\right]\end{aligned}$$

Since $1 - x \leq -\ln x$ for $x \in (0,1]$, it is a standard calculation to see that

$$\operatorname{tr}\left[I - L^T L\right] \leq -\ln\det\left(L^T L\right)$$

We have established:

Corollary 5.24 *Let w be Gaussian white noise at $\{-k, \cdots, k\}$, and zero elsewhere. Then*

$$\sum_{j=-k}^{k} Q(G_{[k,k]})_j \leq 2\gamma^2 \ln \mathcal{E} \exp\Big(\frac{1}{2\gamma^2} \sum_{j=-k}^{k} |z_j|^2\Big)$$
$$= \sum_{j=-k}^{k} E(G_{[k,k]}, \gamma)_j$$

where the output is considered also on the same interval $\{-k, \cdots, k\}$. □

It follows that on a finite horizon the results for the time-varying case mirror the time-invariant results, and the entropy can be interpreted as an LQG cost for which large deviations are penalized.

Remark 5.25 This result shown here is stronger than that of [40] for time-invariant systems, which only deals with the infinite horizon case. This is a result that arises from our time-domain interpretation but that is not available using the usual integral definition of the entropy, which relies on the transfer function of the system.

By relaxing the constraint on γ, that is, allowing γ to tend to infinity, Lemma 5.10 tells us that the entropy operator converges to the quadratic cost operator. Hence all three cost functionals in Corollary 5.24 become equal.

6 Minimum Entropy Control

In this chapter we provide results for the minimum entropy optimal control problem for linear discrete-time time-varying systems. Our approach breaks down to the analysis of a series of simpler control problems, along the line of [31] from which the general result follows. In fact, we exhibit the same separation principle in the entropy as was seen in [53].

First we will give some basic results that we will use throughout the chapter. Then we start with systems for which full information is available to the controller. These results follow directly from a characterization of all sub-optimal controllers. Secondly we consider its dual, the full control problem. A difficulty arises in entropy minimization that does not arise in norm optimization. For a causal operator its norm equals the norm of its anti-causal adjoint, which we saw in Section 4.4 is *not* true for the entropy of a system. Therefore we will minimize the average entropy, from which we have seen in Section 5.3.2 that it mimics the above LTI property.

In the entropy optimization problem for the disturbance feedforward case, the same problem can be overcome, resulting in a different approach from the LTI case. Its dual, the output estimation problem, can be used to solve the general output feedback using a separation principle. The complete solution to the minimum entropy control problem requires solving only two Riccati operator equations.

6.1 Problem statement

Throughout this chapter we will consider a plant $\boldsymbol{G}$ given by the following state-space realization:

$$\Sigma_G := \begin{cases} \boldsymbol{Z}\boldsymbol{x} &= \boldsymbol{A}\boldsymbol{x} + \boldsymbol{B}_1\boldsymbol{w} + \boldsymbol{B}_2\boldsymbol{u} \\ \boldsymbol{z} &= \boldsymbol{C}_1\boldsymbol{x} + \boldsymbol{D}_{11}\boldsymbol{w} + \boldsymbol{D}_{12}\boldsymbol{u} \\ \boldsymbol{y} &= \boldsymbol{C}_2\boldsymbol{x} + \boldsymbol{D}_{21}\boldsymbol{w} \end{cases} \tag{6.1}$$

In (6.1), $\boldsymbol{x} \in \ell_2^n$ is the state of the system. The input $\boldsymbol{w} \in \ell_2^{m_1}$ includes any external inputs as well as any unknown disturbances; $\boldsymbol{z} \in \ell_2^{p_1}$ is the output signal to be controlled; the signal $\boldsymbol{u} \in \ell_2^{m_2}$ is the control input, which can be chosen based on the measurement output $\boldsymbol{y} \in \ell_2^{p_2}$. In terms of the state-space operators in (6.1), describing the system, we make the following assumptions:

A(1) $\boldsymbol{D}_{12}^*\boldsymbol{D}_{12} > 0$

A(2) $\boldsymbol{D}_{21}\boldsymbol{D}_{21}^* > 0$

A(3) The pair $(\boldsymbol{A}, \boldsymbol{B}_2)$ is uniformly stabilizable

A(4) The pair $(\boldsymbol{C}_2, \boldsymbol{A})$ is uniformly detectable

Under the assumption that *A(i)* and *A(2)* are satisfied, the operators $\boldsymbol{D}_{21}^\dagger$ and $\boldsymbol{D}_{12}^\dagger$ given by

$$\boldsymbol{D}_{21}^\dagger := \boldsymbol{D}_{21}^*(\boldsymbol{D}_{21}\boldsymbol{D}_{21}^*)^{-1}$$
$$\boldsymbol{D}_{12}^\dagger := (\boldsymbol{D}_{12}^*\boldsymbol{D}_{12})^{-1}\boldsymbol{D}_{12}^*$$

are well-defined. Using these operators, we make the following additional assumptions:

A(5) The pair $\big((\boldsymbol{I} - \boldsymbol{D}_{12}\boldsymbol{D}_{12}^\dagger)\boldsymbol{C}_1, \boldsymbol{A} - \boldsymbol{B}_2\boldsymbol{D}_{12}^\dagger\boldsymbol{C}_1\big)$ is uniformly detectable

A(6) The pair $\big(\boldsymbol{A} - \boldsymbol{B}_1\boldsymbol{D}_{21}^\dagger\boldsymbol{C}_2, \boldsymbol{B}_1(\boldsymbol{I} - \boldsymbol{D}_{21}^\dagger\boldsymbol{D}_{21})\big)$ is uniformly stabilizable

Although some of the assumptions could be relaxed, they are fairly standard in $\mathcal{H}_\infty$ control theory.

In order to simplify some of the equations, we will use throughout the chapter the notation

$$\boldsymbol{B} := \begin{bmatrix} \boldsymbol{B}_1 & \boldsymbol{B}_2 \end{bmatrix} \qquad \boldsymbol{D}_{1\bullet} := \begin{bmatrix} \boldsymbol{D}_{11} & \boldsymbol{D}_{12} \end{bmatrix}$$

$$\boldsymbol{C} := \begin{bmatrix} \boldsymbol{C}_1 \\ \boldsymbol{C}_2 \end{bmatrix} \qquad \boldsymbol{D}_{\bullet 1} := \begin{bmatrix} \boldsymbol{D}_{11} \\ \boldsymbol{D}_{21} \end{bmatrix}$$

The minimum entropy problem can be stated as follows:

Given $\gamma > 0$, find necessary and sufficient conditions for the existence of a stabilizing controller $\boldsymbol{K}$ such that $\|\mathcal{F}_\ell(\boldsymbol{G}, \boldsymbol{K})\| < \gamma$. If such a controller exists, find the one that minimizes the entropy of the closed-loop system.

Throughout the chapter we will assume that γ equals 1. The results for $\gamma \neq 1$ can always be obtained by scaling the state-space operators; for example in the general output feedback case we can scale $\gamma^{-1/2}\boldsymbol{B}_1$, $\gamma^{1/2}\boldsymbol{B}_2$, $\gamma^{-1/2}\boldsymbol{C}_1$, $\gamma^{1/2}\boldsymbol{C}_2$, $\gamma^{-1}\boldsymbol{D}_{11}$, and get a compensator $\gamma^{-1}\boldsymbol{K}$.

To solve the minimum entropy problem, we will first consider a series of special cases. The method of solution was introduced to the $\mathcal{H}_\infty$ control problem in [31]. The solutions to these problems enable us to solve the general output feedback problem using a separation principle. For the minimum entropy control problem, this separation principle was used for time-invariant systems in [64, 53].

In the next section we will first emphasize the importance of the minimum entropy controller as a trade off between the robustness and the $\mathcal{H}_2$ performance. We will also present some useful preliminary results, which will be used in solving the minimum entropy control problem.

6.2 Basic results

Before presenting some preliminary results, we will first examine one of the most important properties of the minimum entropy controller. In Lemma 5.8 we showed that the entropy operator is an upper bound for the quadratic cost operator. The following lemma shows that the minimum entropy control cost allows one to trade off between the robustness, and the performance of the closed-loop system.

Lemma 6.1 *Suppose that $\gamma \geq \bar{\gamma}$, and denote by $\boldsymbol{G}$ and $\bar{\boldsymbol{G}}$ the closed-loop system with minimum entropy corresponding to γ and $\bar{\gamma}$ respectively. Then $\boldsymbol{E}(\boldsymbol{G}, \gamma) \leq \boldsymbol{E}(\bar{\boldsymbol{G}}, \bar{\gamma})$.*

Proof: We know that

$$\|\bar{\boldsymbol{G}}\| < \bar{\gamma} \leq \gamma$$

Hence $\bar{\boldsymbol{G}}$ also satisfies the norm bound $\|\bar{\boldsymbol{G}}\| < \gamma$. Since $\boldsymbol{G}$ is the closed-loop system minimizing the entropy for γ, it follows that

$$\boldsymbol{E}(\boldsymbol{G}, \gamma) \leq \boldsymbol{E}(\bar{\boldsymbol{G}}, \gamma)$$

Lemma 4.7 tells us that the entropy is a decreasing function with respect to γ, hence

$$\boldsymbol{E}(\bar{\boldsymbol{G}},\gamma) \le \boldsymbol{E}(\bar{\boldsymbol{G}},\bar{\gamma})$$

Combining these two inequalities gives the result. ∎

This result shows that, by increasing γ and permitting a lower level of robustness, the entropy decreases. Since the entropy is an upper bound for the quadratic cost, this ensures that the guaranteed performance of the closed-loop system is better.

Remark 6.2 From the statement above, it would be tempting to conclude that the actual $\mathcal{H}_2$ cost of the closed-loop system will also decrease. In fact, this was originally conjectured in [64]. Mustafa and Glover also conjectured that as the value of γ increased, the achieved $\mathcal{H}_\infty$ norm also increased. Aoki *et al.* presented an example of a LTI system in [5] where the $\mathcal{H}_\infty$ norm of the closed-loop system with a minimum entropy constraint is *not* a decreasing function of γ, contradicting the latter conjecture. In fact, using the same system from [5], it can be shown that the former conjecture is also fallacious. The best that one can say is that the upper bound on the $\mathcal{H}_2$ cost is decreasing with respect to γ.

We now present some basic results that will be used throughout this chapter. The first, usually known as Redheffer's lemma, provides a means of characterizing the set of all closed-loop systems that are contractive. The original result is from [68].

Lemma 6.3 *Suppose that* $\boldsymbol{P} = \left[\begin{smallmatrix} \boldsymbol{P}_{11} & \boldsymbol{P}_{12} \\ \boldsymbol{P}_{21} & \boldsymbol{P}_{22} \end{smallmatrix}\right]$, *with* $\boldsymbol{P}_{11}$, $\boldsymbol{P}_{12}$, $\boldsymbol{P}_{22} \in \mathcal{C}$, *and* $\boldsymbol{P}_{21} \in \mathcal{C} \cap \mathcal{C}^{-1}$, *is an inner operator, i.e.* $\boldsymbol{P}^*\boldsymbol{P} = \boldsymbol{I}$. *Furthermore, assume that* $\boldsymbol{Q}$ *is a causal (not necessarily bounded) operator. The following two statements are equivalent:*

(i) The system is stable and well-posed with $\|\mathcal{F}_\ell(\boldsymbol{P},\boldsymbol{Q})\| < 1$

(ii) $\boldsymbol{Q} \in \mathcal{C}$ *and* $\|\boldsymbol{Q}\| < 1$

Proof: (i)⇐(ii) For our proof, we modify the proof for the time-invariant case found in [31, Lemma 15]).

Since $\boldsymbol{P}$ is inner, $\|\boldsymbol{P}_{22}\| \le 1$. This, together with the fact that $\boldsymbol{Q}$ is a contraction, implies that $\|\boldsymbol{P}_{22}\boldsymbol{Q}\| < 1$. Thus, the series

$$\sum_{k=0}^{\infty}(\boldsymbol{P}_{22}\boldsymbol{Q})^k$$

converges in $\mathcal{C}$ and is equal to $(\boldsymbol{I}-\boldsymbol{P}_{22}\boldsymbol{Q})^{-1}$. This implies that $\boldsymbol{Q}$ stabilizes $\boldsymbol{P}_{22}$. By using a doubly coprime factorization of $\boldsymbol{P}$ (such factorizations are known to exist [24]) and a time-varying version of Lemma 4.2.1 in [37], it follows that $\boldsymbol{Q}$ internally stabilizes $\boldsymbol{P}$. Now to show that $\mathcal{F}_\ell(\boldsymbol{P},\boldsymbol{Q})$ is a contraction, we use the fact that $\boldsymbol{P}$ is inner and a little algebra to get

$$
\begin{aligned}
&\mathcal{F}_\ell(\boldsymbol{P},\boldsymbol{Q})^*\mathcal{F}_\ell(\boldsymbol{P},\boldsymbol{Q}) \\
&\qquad = \boldsymbol{I}-\boldsymbol{P}_{21}^*(\boldsymbol{I}-\boldsymbol{Q}^*\boldsymbol{P}_{22}^*)^{-1}(\boldsymbol{I}-\boldsymbol{Q}^*\boldsymbol{Q})(\boldsymbol{I}-\boldsymbol{P}_{22}\boldsymbol{Q})^{-1}\boldsymbol{P}_{21} \le \boldsymbol{I}
\end{aligned}
$$

where we have used the fact that $\boldsymbol{Q}$ is a contraction.

(ii)⇐(i) To show the converse, we first prove that $\boldsymbol{Q}$ is a bounded operator. Take a right coprime factorization $\boldsymbol{Q}=\boldsymbol{N}\boldsymbol{D}^{-1}$ where $\boldsymbol{N}$, $\boldsymbol{D}\in\mathcal{C}$. Note that $\boldsymbol{Q}\in\mathcal{C}$ if and only if $\boldsymbol{D}\in\mathcal{C}\cap\mathcal{C}^{-1}$; see [35, page 182]. From the internal stability assumption, we know that

$$
\boldsymbol{Q}(\boldsymbol{I}-\boldsymbol{P}_{22}\boldsymbol{Q})^{-1}\in\mathcal{C} \Rightarrow \boldsymbol{N}(\boldsymbol{D}-\boldsymbol{P}_{22}\boldsymbol{N})^{-1}\in\mathcal{C}
$$

Now, since $\boldsymbol{N}$ and $\boldsymbol{D}$ are right coprime, it follows that $\boldsymbol{N}$ and $\boldsymbol{D}-\boldsymbol{P}_{22}\boldsymbol{N}$ are also right coprime. To see this, suppose that $\tilde{\boldsymbol{X}}$, $\tilde{\boldsymbol{Y}}\in\mathcal{C}$ are such that $\tilde{\boldsymbol{X}}\boldsymbol{N}+\tilde{\boldsymbol{Y}}\boldsymbol{D}=\boldsymbol{I}$. Then

$$
\boldsymbol{X}\boldsymbol{N}+\boldsymbol{Y}(\boldsymbol{D}-\boldsymbol{P}_{22}\boldsymbol{N})=\boldsymbol{I}
$$

with $\boldsymbol{X}=\tilde{\boldsymbol{X}}+\boldsymbol{Y}\boldsymbol{P}_{22}\in\mathcal{C}$ and $\boldsymbol{Y}=\tilde{\boldsymbol{Y}}\in\mathcal{C}$ which proves coprimeness. It follows, again from [35, page 182], that

$$
\begin{aligned}
\boldsymbol{N}(\boldsymbol{D}-\boldsymbol{P}_{22}\boldsymbol{N})^{-1}\in\mathcal{C} &\Rightarrow (\boldsymbol{D}-\boldsymbol{P}_{22}\boldsymbol{N})^{-1}\in\mathcal{C} \\
&\Rightarrow \boldsymbol{D}^{-1}(\boldsymbol{I}-\boldsymbol{P}_{22}\boldsymbol{Q})^{-1}\in\mathcal{C} \\
&\Rightarrow \boldsymbol{D}^{-1}\in\mathcal{C}
\end{aligned}
$$

where the last line comes from that fact that $(\boldsymbol{I}-\boldsymbol{P}_{22}\boldsymbol{Q})^{-1}\in\mathcal{C}\cap\mathcal{C}^{-1}$. Thus, $\boldsymbol{Q}\in\mathcal{C}$.

We now show that $\boldsymbol{Q}$ is a contraction. Assume otherwise, thus there exists a signal $\boldsymbol{y}\in\ell_2$ such that $\boldsymbol{u}=\boldsymbol{Q}\boldsymbol{y}\in\ell_2$ and $\|\boldsymbol{u}\|_2\ge\|\boldsymbol{y}\|_2$. Let, $\boldsymbol{w}=\boldsymbol{P}_{21}^{-1}(\boldsymbol{I}-\boldsymbol{P}_{22}\boldsymbol{Q})\boldsymbol{y}$. This is in ℓ_2, since $\boldsymbol{P}_{21}\in\mathcal{C}\cap\mathcal{C}^{-1}$. Moreover, from the inner condition we know that

$$
\begin{aligned}
\|\boldsymbol{z}\|_2^2+\|\boldsymbol{y}\|_2^2 &= \|\boldsymbol{w}\|_2^2+\|\boldsymbol{u}\|_2^2 \\
&\ge \|\boldsymbol{w}\|_2^2+\|\boldsymbol{y}\|_2^2
\end{aligned}
$$

Thus $\|\boldsymbol{z}\|^2\ge\|\boldsymbol{w}\|^2$ which contradicts the assumption that $\mathcal{F}_\ell(\boldsymbol{P},\boldsymbol{Q})$ is a contraction. ■

The following theorem will also be used, and is the essential result used throughout the chapter.

Theorem 6.4 *Suppose that $\mathcal{F}_\ell(\boldsymbol{P},\boldsymbol{Q})$ denotes the set of all closed-loop systems, where $\boldsymbol{P}$ is as in Lemma 6.3, and $\boldsymbol{Q}$ is a causal (bounded) contractive operator. In addition assume that $\boldsymbol{P}_{22}$ is strictly causal. Then*

(i) $\boldsymbol{E}(\mathcal{F}_\ell(\boldsymbol{P},\boldsymbol{Q}))$ is minimized by the unique choice $\boldsymbol{Q}_{\min} := \boldsymbol{0}$

(ii) The closed-loop system with minimum entropy is given by $\boldsymbol{P}_{11}$

Proof: Since $\boldsymbol{Q}$ is a contractive operator, the bounded, Hermitian operator $\boldsymbol{I}-\boldsymbol{Q}^*\boldsymbol{Q}$ has a spectral factorization. Denote the spectral factor by $\boldsymbol{L}$. Now

$$\begin{aligned}
&\boldsymbol{I}-\mathcal{F}_\ell(\boldsymbol{P},\boldsymbol{Q})^*\mathcal{F}_\ell(\boldsymbol{P},\boldsymbol{Q})\\
&\quad= \boldsymbol{P}_{21}^*\left(\boldsymbol{I}-\boldsymbol{Q}^*\boldsymbol{P}_{22}^*\right)^{-1}\left(\boldsymbol{I}-\boldsymbol{Q}^*\boldsymbol{Q}\right)\left(\boldsymbol{I}-\boldsymbol{P}_{22}\boldsymbol{Q}\right)^{-1}\boldsymbol{P}_{21}\\
&\quad= \left[\boldsymbol{L}\left(\boldsymbol{I}-\boldsymbol{P}_{22}\boldsymbol{Q}\right)^{-1}\boldsymbol{P}_{21}\right]^*\left[\boldsymbol{L}\left(\boldsymbol{I}-\boldsymbol{P}_{22}\boldsymbol{Q}\right)^{-1}\boldsymbol{P}_{21}\right]
\end{aligned}$$

We know that $\boldsymbol{P}_{21}\in\mathcal{C}\cap\mathcal{C}^{-1}$. Moreover, $\boldsymbol{L}\in\mathcal{C}\cap\mathcal{C}^{-1}$, since $\boldsymbol{L}$ is a spectral factor. Finally, we have seen in the proof of Lemma 6.3 that the operator $\boldsymbol{I}-\boldsymbol{P}_{22}\boldsymbol{Q}$ is in $\mathcal{C}\cap\mathcal{C}^{-1}$. Thus

$$\boldsymbol{M} := \left[\boldsymbol{L}\left(\boldsymbol{I}-\boldsymbol{P}_{22}\boldsymbol{Q}\right)^{-1}\boldsymbol{P}_{21}\right]$$

is a causal operator with a causal inverse. Hence it is clearly a spectral factor of the operator $\boldsymbol{I}-\mathcal{F}_\ell(\boldsymbol{P},\boldsymbol{Q})^*\mathcal{F}_\ell(\boldsymbol{P},\boldsymbol{Q})$. Since all the operators in $\boldsymbol{M}$ are causal, we get

$$\begin{aligned}
M_{k,k} &= L_{k,k}\left(\left(\boldsymbol{I}-\boldsymbol{P}_{22}\boldsymbol{Q}\right)^{-1}\right)_{k,k}P_{21_{k,k}}\\
&= L_{k,k}\left(\sum_{i=0}^{\infty}(\boldsymbol{P}_{22}\boldsymbol{Q})^i\right)_{k,k}P_{21_{k,k}}\\
&= L_{k,k}\sum_{i=0}^{\infty}\left(P_{22_{k,k}}Q_{k,k}\right)^i P_{21_{k,k}}\\
&= L_{k,k}P_{21_{k,k}}
\end{aligned}$$

since $P_{22_{k,k}}=0$.

It follows that the entropy can be evaluated as

$$\begin{aligned} \boldsymbol{E}(\mathcal{F}_\ell(\boldsymbol{P},\boldsymbol{Q})) &= -\operatorname{diag}\left\{\ln\det\left(P_{21_{k,k}}^T L_{k,k}^T L_{k,k} P_{21_{k,k}}\right)\right\} \\ &= -\operatorname{diag}\left\{\ln\det\left(P_{21_{k,k}}^T P_{21_{k,k}}\right)\right\} - \operatorname{diag}\left\{\ln\det\left(L_{k,k}^T L_{k,k}\right)\right\} \\ &= \boldsymbol{E}(\boldsymbol{P}_{11}) + \boldsymbol{E}(\boldsymbol{Q}) \end{aligned}$$

The last step is a result of $\boldsymbol{P}$ being inner. Namely $\boldsymbol{I} - \boldsymbol{P}_{11}^*\boldsymbol{P}_{11} = \boldsymbol{P}_{21}^*\boldsymbol{P}_{21}$ is a spectral factorization since $\boldsymbol{P}_{21} \in \mathcal{C} \cap \mathcal{C}^{-1}$. The first term is fixed, and it follows from Lemma 4.5 that the second term is minimized uniquely by $\boldsymbol{Q}_{\min} = \boldsymbol{0}$. The closed-loop system with minimum entropy follows immediately from $\mathcal{F}_\ell(\boldsymbol{P},\boldsymbol{0}) = \boldsymbol{P}_{11}$. ■

In the next section we will solve the minimum entropy control problem for the full information case.

6.3 Full information

In the full information (FI) case the measurement output has complete information about both state and disturbance. A way to represent such systems in state-space form is as in (6.1) with

$$\boldsymbol{C}_2 = \begin{bmatrix} \boldsymbol{I} \\ \boldsymbol{0} \end{bmatrix} \quad \text{and} \quad \boldsymbol{D}_{21} = \begin{bmatrix} \boldsymbol{0} \\ \boldsymbol{I} \end{bmatrix}$$

Note that this is not a special case of the general problem. For notational convenience the output $\boldsymbol{y}$ is re-arranged as $\begin{bmatrix} \boldsymbol{x}^T & \boldsymbol{w}^T \end{bmatrix}^T$ instead of a vector which has as k^{th} element $\begin{bmatrix} x_k^T & w_k^T \end{bmatrix}^T$. For the FI system we assume that assumptions *A(1)*, *A(3)*, and *A(5)* hold.

Before solving the minimum entropy control problem for the FI problem, we need to find necessary and sufficient conditions for the existence of a stabilizing controller achieving a norm bound. Whenever these conditions are satisfied, we can parameterize all suitable controllers, and choose the one which minimizes the entropy.

6.3.1 Characterizing all closed-loop systems

For the system $\boldsymbol{G}_{FI}$ we find the following necessary conditions for the existence of a stabilizing controller achieving a prescribed norm bound. Since the proof is quite technical, it is given in Appendix A.

Theorem 6.5 *If there exists a stabilizing controller* $\boldsymbol{K}$ *such that*

$$\boldsymbol{I} - \mathcal{F}_\ell(\boldsymbol{G}_{FI}, \boldsymbol{K})^* \mathcal{F}_\ell(\boldsymbol{G}_{FI}, \boldsymbol{K}) > \boldsymbol{0}$$

then there exist a memoryless operator $\boldsymbol{X} \geq \boldsymbol{0}$ *and*

$$\boldsymbol{T} := \begin{bmatrix} \boldsymbol{T}_{11} & \boldsymbol{0} \\ \boldsymbol{T}_{21} & \boldsymbol{T}_{22} \end{bmatrix}$$

where $\boldsymbol{T}_{11}$, $\boldsymbol{T}_{21}$ *and* $\boldsymbol{T}_{22}$ *are all memoryless operators with* $\boldsymbol{T}_{11} > \boldsymbol{0}$ *and* $\boldsymbol{T}_{22} > \boldsymbol{0}$ *such that*

$$\boldsymbol{R} + \boldsymbol{B}^*\boldsymbol{Z}\boldsymbol{X}\boldsymbol{Z}^*\boldsymbol{B} = \boldsymbol{T}^*\boldsymbol{J}\boldsymbol{T} \tag{6.2}$$

$$\begin{aligned} \boldsymbol{X} = {} & \boldsymbol{A}^*\boldsymbol{Z}\boldsymbol{X}\boldsymbol{Z}^*\boldsymbol{A} + \boldsymbol{C}_1^*\boldsymbol{C}_1 \\ & - (\boldsymbol{A}^*\boldsymbol{Z}\boldsymbol{X}\boldsymbol{Z}^*\boldsymbol{B} + \boldsymbol{C}_1^*\boldsymbol{D}_{1\bullet})(\boldsymbol{R} + \boldsymbol{B}^*\boldsymbol{Z}\boldsymbol{X}\boldsymbol{Z}^*\boldsymbol{B})^{-1} \\ & \times (\boldsymbol{B}^*\boldsymbol{Z}\boldsymbol{X}\boldsymbol{Z}^*\boldsymbol{A} + \boldsymbol{D}_{1\bullet}^*\boldsymbol{C}_1) \end{aligned} \tag{6.3}$$

and

$$\boldsymbol{A}_{cl} := \boldsymbol{A} - \boldsymbol{B}(\boldsymbol{R} + \boldsymbol{B}^*\boldsymbol{Z}\boldsymbol{X}\boldsymbol{Z}^*\boldsymbol{B})^{-1}(\boldsymbol{B}^*\boldsymbol{Z}\boldsymbol{X}\boldsymbol{Z}^*\boldsymbol{A} + \boldsymbol{D}_{1\bullet}^*\boldsymbol{C}_1)$$

is UES, where

$$\boldsymbol{R} := \boldsymbol{D}_{1\bullet}^*\boldsymbol{D}_{1\bullet} + \begin{bmatrix} -\boldsymbol{I} & \boldsymbol{0} \\ \boldsymbol{0} & \boldsymbol{0} \end{bmatrix} \quad \text{and} \quad \boldsymbol{J} := \begin{bmatrix} -\boldsymbol{I} & \boldsymbol{0} \\ \boldsymbol{0} & \boldsymbol{I} \end{bmatrix} \qquad \square$$

As in the proof of Lemma 3.2, it can be shown that the operator Riccati equation (6.3) has a unique stabilizing solution. To show the converse of Theorem 6.5, and also give a parameterization of all stabilizing controllers achieving the norm bound, we introduce the following variables, assuming that the conditions in Theorem 6.5 hold.

$$\begin{aligned} \boldsymbol{v} & := \boldsymbol{T}_{22}\boldsymbol{u} + \boldsymbol{T}_{21}\boldsymbol{w} - \begin{bmatrix} \boldsymbol{T}_{21} & \boldsymbol{T}_{22} \end{bmatrix} \boldsymbol{F}\boldsymbol{x} \\ \boldsymbol{r} & := \boldsymbol{T}_{11}(\boldsymbol{w} - \boldsymbol{F}_1\boldsymbol{x}) \\ \boldsymbol{F} & = \begin{bmatrix} \boldsymbol{F}_1 \\ \boldsymbol{F}_2 \end{bmatrix} := -(\boldsymbol{R} + \boldsymbol{B}^*\boldsymbol{Z}\boldsymbol{X}\boldsymbol{Z}^*\boldsymbol{B})^{-1}(\boldsymbol{B}^*\boldsymbol{Z}\boldsymbol{X}\boldsymbol{Z}^*\boldsymbol{A} + \boldsymbol{D}_{1\bullet}^*\boldsymbol{C}_1) \end{aligned}$$

Regarding $\boldsymbol{r}$ as control input and $\boldsymbol{v}$ as measurement output, we get the system $\boldsymbol{P} = \left[\begin{smallmatrix} \boldsymbol{P}_{11} & \boldsymbol{P}_{12} \\ \boldsymbol{P}_{21} & \boldsymbol{P}_{22} \end{smallmatrix}\right]$, with state-space realization

$$\Sigma_P = \left[\begin{array}{c|c} \boldsymbol{A}_P & \boldsymbol{B}_P \\ \hline \boldsymbol{C}_P & \boldsymbol{D}_P \end{array}\right] := \left[\begin{array}{c|cc} \boldsymbol{A}_F & \boldsymbol{B}_1 - \boldsymbol{B}_2\boldsymbol{T}_{22}^{-1}\boldsymbol{T}_{21} & \boldsymbol{B}_2\boldsymbol{T}_{22}^{-1} \\ \hline \boldsymbol{C}_F & \boldsymbol{D}_{11} - \boldsymbol{D}_{12}\boldsymbol{T}_{22}^{-1}\boldsymbol{T}_{21} & \boldsymbol{D}_{12}\boldsymbol{T}_{22}^{-1} \\ -\boldsymbol{T}_{11}\boldsymbol{F}_1 & \boldsymbol{T}_{11} & \boldsymbol{0} \end{array}\right] \tag{6.4}$$

where

$$\boldsymbol{A}_F := \boldsymbol{A} + \boldsymbol{B}_2 \left[\begin{array}{cc} \boldsymbol{T}_{22}^{-1}\boldsymbol{T}_{21} & \boldsymbol{I} \end{array}\right] \boldsymbol{F}$$
$$\boldsymbol{C}_F := \boldsymbol{C}_1 + \boldsymbol{D}_{12} \left[\begin{array}{cc} \boldsymbol{T}_{22}^{-1}\boldsymbol{T}_{21} & \boldsymbol{I} \end{array}\right] \boldsymbol{F}$$

We will now show that the operator $\boldsymbol{P}$ is inner; i.e. it has a norm preserving property. This will enable us to use the operator version of Redheffer's lemma, Lemma 6.3, to characterize the set of all admissible closed-loop systems for which the desired norm bound holds.

Lemma 6.6 *The operator $\boldsymbol{P}$, given by (6.4), is inner.*

Proof: We want to show that $\boldsymbol{P}^*\boldsymbol{P} = \boldsymbol{I}$. First we show that $\boldsymbol{X}$ is the observability Gramian of $\boldsymbol{P}$. Expanding out the expression (6.2), results in

$$\begin{bmatrix} \boldsymbol{D}_{11}^*\boldsymbol{D}_{11} + \boldsymbol{B}_1^*\boldsymbol{ZXZ}^*\boldsymbol{B}_1 - \boldsymbol{I} & \boldsymbol{D}_{11}^*\boldsymbol{D}_{12} + \boldsymbol{B}_1^*\boldsymbol{ZXZ}^*\boldsymbol{B}_2 \\ \boldsymbol{D}_{12}^*\boldsymbol{D}_{11} + \boldsymbol{B}_2^*\boldsymbol{ZXZ}^*\boldsymbol{B}_1 & \boldsymbol{D}_{12}^*\boldsymbol{D}_{12} + \boldsymbol{B}_2^*\boldsymbol{ZXZ}^*\boldsymbol{B}_2 \end{bmatrix}$$
$$= \begin{bmatrix} \boldsymbol{T}_{21}^*\boldsymbol{T}_{21} - \boldsymbol{T}_{11}^*\boldsymbol{T}_{11} & \boldsymbol{T}_{21}^*\boldsymbol{T}_{22} \\ \boldsymbol{T}_{22}^*\boldsymbol{T}_{21} & \boldsymbol{T}_{22}^*\boldsymbol{T}_{22} \end{bmatrix}$$

Using these expressions and the definition of $\boldsymbol{F}$, we can obtain the identity.

$$-\left[\begin{array}{cc} \boldsymbol{0} & \boldsymbol{I} \end{array}\right](\boldsymbol{R} + \boldsymbol{B}^*\boldsymbol{ZXZ}^*\boldsymbol{B})\boldsymbol{F} = \boldsymbol{B}_2^*\boldsymbol{ZXZ}^*\boldsymbol{A} + \boldsymbol{D}_{12}^*\boldsymbol{C}_1$$

Now, using these two expressions, it is straightforward to show that

$$\begin{aligned}
&\boldsymbol{A}_P^*\boldsymbol{ZXZ}^*\boldsymbol{A}_P + \boldsymbol{C}_P^*\boldsymbol{C}_P \\
&\quad = \boldsymbol{A}_F^*\boldsymbol{ZXZ}^*\boldsymbol{A}_F + \boldsymbol{C}_F^*\boldsymbol{C}_F + \boldsymbol{F}_1^*\boldsymbol{T}_{11}^*\boldsymbol{T}_{11}\boldsymbol{F}_1 \\
&\quad = \boldsymbol{A}^*\boldsymbol{ZXZ}^*\boldsymbol{A} + \boldsymbol{C}_1^*\boldsymbol{C}_1 + (\boldsymbol{A}^*\boldsymbol{ZXZ}^*\boldsymbol{B}_2 + \boldsymbol{C}_1^*\boldsymbol{D}_{12})\left[\begin{array}{cc} \boldsymbol{T}_{22}^{-1}\boldsymbol{T}_{21} & \boldsymbol{I} \end{array}\right]\boldsymbol{F} \\
&\qquad + \boldsymbol{F}^*\begin{bmatrix} \boldsymbol{T}_{21}^*\boldsymbol{T}_{22}^{-*} \\ \boldsymbol{I} \end{bmatrix}(\boldsymbol{B}_2^*\boldsymbol{ZXZ}^*\boldsymbol{A} + \boldsymbol{D}_{12}^*\boldsymbol{C}_1) \\
&\qquad + \boldsymbol{F}^*\begin{bmatrix} \boldsymbol{T}_{21}^*\boldsymbol{T}_{21} + \boldsymbol{T}_{11}^*\boldsymbol{T}_{11} & \boldsymbol{T}_{21}^*\boldsymbol{T}_{22} \\ \boldsymbol{T}_{22}^*\boldsymbol{T}_{21} & \boldsymbol{T}_{22}^*\boldsymbol{T}_{22} \end{bmatrix}\boldsymbol{F} \\
&\quad = \boldsymbol{X}
\end{aligned}$$

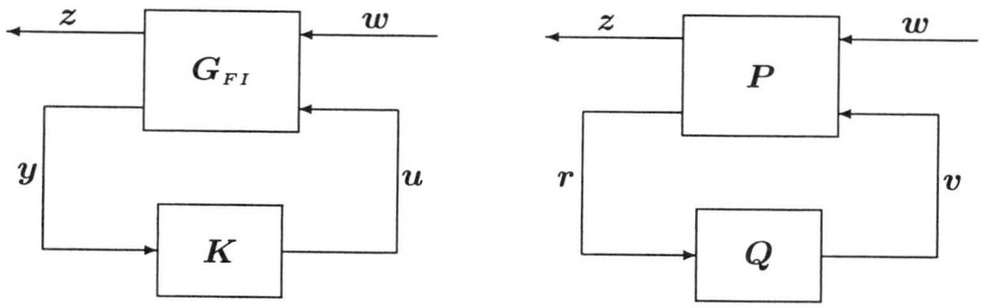

Figure 6.1: $\mathcal{F}_\ell(G_{FI}, K)$ and $\mathcal{F}_\ell(P, Q)$.

Furthermore, an easy calculation shows that

$$\begin{aligned} &D_P^* C_P + B_P^* Z X Z^* A_P \\ &= \begin{bmatrix} I & -T_{21}^* T_{22}^{-*} \\ 0 & T_{22}^{-*} \end{bmatrix} (D_{1\bullet}^* C_1 + B^* Z X Z^* A + (R + B^* Z X Z^* B) F) \\ &= 0 \end{aligned}$$

where the last equality comes from the definition of F. Combining these two equalities enables us to compute $P^* P$ as

$$\begin{aligned} &(D_P + C_P(I - Z^* A_P)^{-1} Z^* B_P)^* (D_P + C_P(I - Z^* A_P)^{-1} Z^* B_P) \\ &= D_P^* D_P + B_P^* Z X Z^* B_P \\ &= \begin{bmatrix} I & -T_{21}^* T_{22}^{-*} \\ 0 & T_{22}^{-*} \end{bmatrix} \begin{bmatrix} D_{11}^* & T_{11}^* \\ D_{12}^* & 0 \end{bmatrix} \begin{bmatrix} D_{11} & D_{12} \\ T_{11} & 0 \end{bmatrix} \begin{bmatrix} I & 0 \\ -T_{22}^{-1} T_{21} & T_{22}^{-1} \end{bmatrix} \\ &\quad + \begin{bmatrix} I & -T_{21}^* T_{22}^{-*} \\ 0 & T_{22}^{-*} \end{bmatrix} B^* Z X Z^* B \begin{bmatrix} I & 0 \\ -T_{22}^{-1} T_{21} & T_{22}^{-1} \end{bmatrix} \\ &= \begin{bmatrix} I & -T_{21}^* T_{22}^{-*} \\ 0 & T_{22}^{-*} \end{bmatrix} \begin{bmatrix} I + T_{21}^* T_{21} & T_{21}^* T_{22} \\ T_{22}^* T_{21} & T_{22}^* T_{22} \end{bmatrix} \begin{bmatrix} I & 0 \\ -T_{22}^{-1} T_{21} & T_{22}^{-1} \end{bmatrix} \\ &= \begin{bmatrix} I & 0 \\ 0 & I \end{bmatrix} \end{aligned}$$

as required. ■

We will now show that the two feedback configurations in Figure 6.1 are equivalent. This, together with Lemma 6.3, allows us to prove the converse of Theorem 6.5.

Theorem 6.7 *If there exist $\boldsymbol{X}$ and $\boldsymbol{T}$ satisfying the conditions in Theorem 6.5, then there exists a stabilizing controller $\boldsymbol{K}$ such that*

$$\boldsymbol{I} - \mathcal{F}_\ell(\boldsymbol{G}_{FI}, \boldsymbol{K})^* \mathcal{F}_\ell(\boldsymbol{G}_{FI}, \boldsymbol{K}) > 0$$

Moreover, all closed-loop systems $\mathcal{F}_\ell(\boldsymbol{G}_{FI}, \boldsymbol{K})$ satisfying the norm constraint $\|\mathcal{F}_\ell(\boldsymbol{G}_{FI}, \boldsymbol{K})\| < 1$ can be written as $\mathcal{F}_\ell(\boldsymbol{P}, \boldsymbol{Q})$, where $\boldsymbol{Q}$ is a causal operator with $\|\boldsymbol{Q}\| < 1$.

Proof: We have seen that $\boldsymbol{X}$ is the observability Gramian of $\boldsymbol{P}$. Since $\boldsymbol{A}_F + (\boldsymbol{B}_1 - \boldsymbol{B}_2\boldsymbol{T}_{22}^{-1}\boldsymbol{T}_{21})\boldsymbol{F}_1 = \boldsymbol{A}_{cl}$, and this is UES, it follows that $(\boldsymbol{C}_P, \boldsymbol{A}_P)$ is uniformly detectable, and hence $\boldsymbol{A}_P$ is UES [2]. Obviously the elements of $\boldsymbol{P}$ are in $\mathcal{C}$. Since

$$\boldsymbol{P}_{21}^{-1} = \boldsymbol{T}_{11}^{-1} + \boldsymbol{F}_1\Big(\boldsymbol{I} - \boldsymbol{Z}^*(\boldsymbol{A} + \boldsymbol{B}\boldsymbol{F})\Big)^{-1}\boldsymbol{Z}^*(\boldsymbol{B}_1 - \boldsymbol{B}_2\boldsymbol{T}_{22}^{-1}\boldsymbol{T}_{21})\boldsymbol{T}_{11}^{-1}$$

we also have $\boldsymbol{P}_{21}^{-1} \in \mathcal{C}$. It follows from Lemma 6.3 and the definitions of $\boldsymbol{v}$ and $\boldsymbol{r}$ that, if $\boldsymbol{Q}$ is any causal operator with $\|\boldsymbol{Q}\| < 1$, then the compensator

$$\boldsymbol{K} = \boldsymbol{T}_{22}^{-1}\begin{bmatrix} -\boldsymbol{Q} & \boldsymbol{I} \end{bmatrix}\begin{bmatrix} \boldsymbol{T}_{11} & \boldsymbol{0} \\ \boldsymbol{T}_{21} & \boldsymbol{T}_{22} \end{bmatrix}\begin{bmatrix} \boldsymbol{F}_1 & -\boldsymbol{I} \\ \boldsymbol{F}_2 & \boldsymbol{0} \end{bmatrix} \tag{6.5}$$

is stabilizing, and $\|\mathcal{F}_\ell(\boldsymbol{G}_{FI}, \boldsymbol{K})\| < 1$. On the other hand, if $\boldsymbol{K}$ is a stabilizing controller achieving the norm bound, then the operator $\boldsymbol{T}_{vw}$ mapping $\boldsymbol{w}$ to $\boldsymbol{v}$ is also stable. Since $\boldsymbol{P}_{22}$ is strictly proper, the operator $\boldsymbol{Q} := (\boldsymbol{I} + \boldsymbol{T}_{vw}\boldsymbol{P}_{21}^{-1}\boldsymbol{P}_{22})^{-1}\boldsymbol{T}_{vw}\boldsymbol{P}_{21}^{-1}$ is well-defined. Furthermore the operator from $\boldsymbol{w}$ to $\boldsymbol{z}$ satisfies

$$\begin{aligned} \boldsymbol{T}_{zw} &= \boldsymbol{P}_{11} + \boldsymbol{P}_{12}\boldsymbol{T}_{vw} \\ &= \boldsymbol{P}_{11} + \boldsymbol{P}_{12}\boldsymbol{Q}(\boldsymbol{I} - \boldsymbol{P}_{22}\boldsymbol{Q})^{-1}\boldsymbol{P}_{21} \\ &= \mathcal{F}_\ell(\boldsymbol{P}, \boldsymbol{Q}) \end{aligned}$$

It follows from Lemma 6.3 that $\|\boldsymbol{Q}\| < 1$, which shows that all stable closed-loop system satisfying the norm bound can be characterized by $\mathcal{F}_\ell(\boldsymbol{P}, \boldsymbol{Q})$ for $\boldsymbol{Q} \in \mathcal{C}$, and $\|\boldsymbol{Q}\| < 1$. ■

6.3.2 FI minimum entropy controller

We are now ready to select the entropy minimizing controller from the set (6.5). We have shown that all closed-loop systems are parameterized by the set $\mathcal{F}_\ell(\boldsymbol{P}, \boldsymbol{Q})$ where $\|\boldsymbol{Q}\| < 1$. This allows us to use Theorem 6.4 to choose the controller which minimizes the entropy.

Lemma 6.8 *The entropy is minimized by the controller*

$$\boldsymbol{K}_{\min} = \boldsymbol{T}_{22}^{-1} \begin{bmatrix} \boldsymbol{T}_{21} & \boldsymbol{T}_{22} \end{bmatrix} \begin{bmatrix} \boldsymbol{F}_1 & -\boldsymbol{I} \\ \boldsymbol{F}_2 & \boldsymbol{0} \end{bmatrix}$$

and the minimum value of the entropy is given by

$$\boldsymbol{E}\big(\mathcal{F}_\ell(\boldsymbol{G}_{FI}, \boldsymbol{K}_{\min})\big) = \boldsymbol{E}(\boldsymbol{P}_{11}) = -\operatorname{diag}\left\{\ln\det\left(T_{11_k}^T T_{11_k}\right)\right\} \tag{6.6}$$

Proof: It follows from Theorem 6.7 and Theorem 6.4 that $\boldsymbol{Q}_{\min} = \boldsymbol{0}$ gives the closed-loop system with minimum entropy. The compensator $\boldsymbol{K}_{\min}$ follows by substituting $\boldsymbol{Q}_{\min} = \boldsymbol{0}$ in (6.5). Since $\boldsymbol{I} - \boldsymbol{P}_{11}^*\boldsymbol{P}_{11} = \boldsymbol{P}_{21}^*\boldsymbol{P}_{21}$, the minimum value of the entropy can be written as

$$\begin{aligned} \boldsymbol{E}\big(\mathcal{F}_\ell(\boldsymbol{G}_{FI}, \boldsymbol{K}_{\min})\big) &= \boldsymbol{E}\big(\mathcal{F}_\ell(\boldsymbol{P}, \boldsymbol{Q}_{\min})\big) \\ &= \boldsymbol{E}(\boldsymbol{P}_{11}) \\ &= -\operatorname{diag}\left\{\ln\det\left(P_{21_{k,k}}^T P_{21_{k,k}}\right)\right\} \\ &= -\operatorname{diag}\left\{\ln\det\left(T_{11_k}^T T_{11_k}\right)\right\} \end{aligned}$$

which also follows from Theorem 6.4. ■

Remark 6.9 The minimum value of the entropy can be expressed in terms of the operators of the plant and the solution to the operator Riccati equation (6.3). According to the proof of Claim 13 in Appendix A, this yields

$$\begin{aligned} \boldsymbol{E}\big(\mathcal{F}_\ell(\boldsymbol{G}_{FI}, \boldsymbol{K}_{\min})\big) = -\operatorname{diag}\big\{\ln\det\big(&I - D_{11_k}^T D_{11_k} - B_{1_k}^T X_{k+1} B_{1_k} \\ &+ R_k^T (D_{12_k}^T D_{12_k} + B_{2_k}^T X_{k+1} B_{2_k})^{-1} R_k\big)\big\} \end{aligned}$$

where $R_k = B_{2_k}^T X_{k+1} B_{1_k} + D_{12_k}^T D_{11_k}$.

In the next section we will consider the minimum entropy control problem for the dual of the full information case, the so-called full control problem.

6.4 Full control

In the time-invariant case the results for the dual of the FI problem, the full control (FC) problem, can be obtained directly from the FI results. In the time-varying case however, this is not the case, due to the fact that the entropy of a causal system does not equal the entropy of its (anticausal) adjoint. Nevertheless, the results for the FI case can still be used to parameterize the set of suitable closed-loop systems, by using the concept of a natural dual.

The full control problem relates to systems of the form (6.1) with

$$\boldsymbol{B}_2 = \begin{bmatrix} \boldsymbol{I} & \boldsymbol{0} \end{bmatrix} \quad \text{and} \quad \boldsymbol{D}_{12} = \begin{bmatrix} \boldsymbol{0} & \boldsymbol{I} \end{bmatrix}$$

For notational convenience the input $\boldsymbol{u}$ is re-arranged such that the corresponding operators have the given form. The assumptions *A(2)*, *A(4)*, and *A(6)* are valid for this problem.

The FC problem is related to the FI case, and we can use the results of the FI case to find necessary and sufficient conditions for the existence of a stabilizing compensator achieving a norm bound on the closed-loop system.

Theorem 6.10 *There exists a stabilizing controller $\boldsymbol{K}$ such that*

$$\boldsymbol{I} - \mathcal{F}_\ell(\boldsymbol{G}_{FC}, \boldsymbol{K})^* \mathcal{F}_\ell(\boldsymbol{G}_{FC}, \boldsymbol{K}) > \boldsymbol{0}$$

if and only if there exist a memoryless operator $\boldsymbol{Y} \geq \boldsymbol{0}$ and

$$\tilde{\boldsymbol{T}} := \begin{bmatrix} \tilde{\boldsymbol{T}}_{11} & \tilde{\boldsymbol{T}}_{12} \\ \boldsymbol{0} & \tilde{\boldsymbol{T}}_{22} \end{bmatrix}$$

where $\tilde{\boldsymbol{T}}_{11}$, $\tilde{\boldsymbol{T}}_{12}$ and $\tilde{\boldsymbol{T}}_{22}$ are all memoryless operators with $\tilde{\boldsymbol{T}}_{11} > \boldsymbol{0}$ and $\tilde{\boldsymbol{T}}_{22} > \boldsymbol{0}$ such that

$$\tilde{\boldsymbol{R}} + \boldsymbol{C}\boldsymbol{Z}^*\boldsymbol{Y}\boldsymbol{Z}\boldsymbol{C}^* = \tilde{\boldsymbol{T}}\boldsymbol{J}\tilde{\boldsymbol{T}}^*$$

$$\begin{aligned} \boldsymbol{Y} = {} & \boldsymbol{A}\boldsymbol{Z}^*\boldsymbol{Y}\boldsymbol{Z}\boldsymbol{A}^* + \boldsymbol{B}_1\boldsymbol{B}_1^* \\ & - (\boldsymbol{A}\boldsymbol{Z}^*\boldsymbol{Y}\boldsymbol{Z}\boldsymbol{C}^* + \boldsymbol{B}_1\boldsymbol{D}_{\bullet 1}^*)(\tilde{\boldsymbol{R}} + \boldsymbol{C}\boldsymbol{Z}^*\boldsymbol{Y}\boldsymbol{Z}\boldsymbol{C}^*)^{-1} \\ & \times (\boldsymbol{C}\boldsymbol{Z}^*\boldsymbol{Y}\boldsymbol{Z}\boldsymbol{A}^* + \boldsymbol{D}_{\bullet 1}\boldsymbol{B}_1^*) \end{aligned} \tag{6.7}$$

and

$$\boldsymbol{A}_{cl} := \boldsymbol{A} - (\boldsymbol{A}\boldsymbol{Z}^*\boldsymbol{Y}\boldsymbol{Z}\boldsymbol{C}^* + \boldsymbol{B}_1\boldsymbol{D}_{\bullet 1}^*)(\tilde{\boldsymbol{R}} + \boldsymbol{C}\boldsymbol{Z}^*\boldsymbol{Y}\boldsymbol{Z}\boldsymbol{C}^*)^{-1}\boldsymbol{C}$$

is UES, where

$$\tilde{\boldsymbol{R}} := \boldsymbol{D}_{\bullet 1}\boldsymbol{D}_{\bullet 1}^* + \begin{bmatrix} -\boldsymbol{I} & 0 \\ 0 & 0 \end{bmatrix} \quad \textit{and} \quad \boldsymbol{J} := \begin{bmatrix} -\boldsymbol{I} & 0 \\ 0 & \boldsymbol{I} \end{bmatrix}$$

Proof: First we look at the natural dual $\Omega \boldsymbol{G}_{FC}^* \Omega$ of $\boldsymbol{G}_{FC}$, as defined in Section 2.3. Lemma 2.13 tells us that this has a state-space realization

$$\Sigma_{\Omega G_{FC}^*\Omega} = \left\{ \begin{array}{rcl} \boldsymbol{Z}\boldsymbol{x} & = & \Omega\boldsymbol{A}^*\Omega\boldsymbol{x} + \Omega\boldsymbol{C}_1^*\Omega\boldsymbol{w} + \Omega\boldsymbol{C}_2^*\Omega\boldsymbol{u} \\ \boldsymbol{z} & = & \Omega\boldsymbol{B}_1^*\Omega\boldsymbol{x} + \Omega\boldsymbol{D}_{11}^*\Omega\boldsymbol{w} + \Omega\boldsymbol{D}_{21}^*\Omega\boldsymbol{u} \\ \boldsymbol{y} & = & \Omega\begin{bmatrix} \boldsymbol{I} \\ \boldsymbol{0} \end{bmatrix}\Omega\boldsymbol{x} + \Omega\begin{bmatrix} \boldsymbol{0} \\ \boldsymbol{I} \end{bmatrix}\Omega\boldsymbol{w} \end{array} \right.$$

Since $\Omega = \Omega^*$ and $\Omega^2 = \boldsymbol{I}$, Lemma 2.14 tells us that $\|\mathcal{F}_\ell(\boldsymbol{G}_{FC}, \boldsymbol{K})\| < 1$ if and only if $\|\mathcal{F}_\ell(\Omega \boldsymbol{G}_{FC}^* \Omega, \Omega \boldsymbol{K}^* \Omega)\| < 1$. Using Lemma 2.12 it is immediate that the system $\Omega \boldsymbol{G}_{FC}^* \Omega$ satisfies the conditions for the FI problem, so we can apply Theorems 6.5 and 6.7 to write down the Riccati operator equation corresponding to this system.

$$\begin{aligned}
\boldsymbol{X} &= \Omega \boldsymbol{A} \Omega \boldsymbol{Z} \boldsymbol{X} \boldsymbol{Z}^* \Omega \boldsymbol{A}^* \Omega + \Omega \boldsymbol{B}_1 \boldsymbol{B}_1^* \Omega \\
&\quad - \boldsymbol{L}^* \left(\boldsymbol{R} + \begin{bmatrix} \Omega & \boldsymbol{0} \\ \boldsymbol{0} & \Omega \end{bmatrix} \boldsymbol{C} \Omega \boldsymbol{Z} \boldsymbol{X} \boldsymbol{Z}^* \Omega \boldsymbol{C}^* \begin{bmatrix} \Omega & \boldsymbol{0} \\ \boldsymbol{0} & \Omega \end{bmatrix} \right)^{-1} \boldsymbol{L} \\
\boldsymbol{L} &= \begin{bmatrix} \Omega & \boldsymbol{0} \\ \boldsymbol{0} & \Omega \end{bmatrix} \boldsymbol{C} \Omega \boldsymbol{Z} \boldsymbol{X} \boldsymbol{Z}^* \Omega \boldsymbol{A}^* \Omega + \begin{bmatrix} \Omega & \boldsymbol{0} \\ \boldsymbol{0} & \Omega \end{bmatrix} \boldsymbol{D}_{\bullet 1} \boldsymbol{B}_1^* \Omega \\
\boldsymbol{R} &= \begin{bmatrix} \Omega & \boldsymbol{0} \\ \boldsymbol{0} & \Omega \end{bmatrix} \left(\boldsymbol{D}_{\bullet 1} \boldsymbol{D}_{\bullet 1}^* + \begin{bmatrix} -\boldsymbol{I} & \boldsymbol{0} \\ \boldsymbol{0} & \boldsymbol{0} \end{bmatrix} \right) \begin{bmatrix} \Omega & \boldsymbol{0} \\ \boldsymbol{0} & \Omega \end{bmatrix}
\end{aligned}$$

Note that we used repeatedly the fact that $\Omega^2 = \boldsymbol{I}$. Since $\Omega \boldsymbol{Z} \Omega = \boldsymbol{Z}^*$, the Riccati operator equation (6.7) follows by pre- and post-multiplying with Ω and setting $\boldsymbol{Y} = \Omega \boldsymbol{X} \Omega$. The FI theorem gives a closed-loop matrix $\boldsymbol{A}_{clX}$ which is UES, and since $\boldsymbol{A}_{cl} = \Omega \boldsymbol{A}_{clX}^* \Omega$ the stability result for $\boldsymbol{A}_{cl}$ follows from Lemma 2.12. Furthermore we know that

$$\boldsymbol{R} + \begin{bmatrix} \Omega & \boldsymbol{0} \\ \boldsymbol{0} & \Omega \end{bmatrix} \boldsymbol{C} \Omega \boldsymbol{Z} \boldsymbol{X} \boldsymbol{Z}^* \Omega \boldsymbol{C}^* \begin{bmatrix} \Omega & \boldsymbol{0} \\ \boldsymbol{0} & \Omega \end{bmatrix} = \boldsymbol{T}^* \boldsymbol{J} \boldsymbol{T}$$

where

$$\boldsymbol{T} = \begin{bmatrix} \boldsymbol{T}_{11} & \boldsymbol{0} \\ \boldsymbol{T}_{21} & \boldsymbol{T}_{22} \end{bmatrix}$$

$\boldsymbol{T}_{11}$, $\boldsymbol{T}_{21}$ and $\boldsymbol{T}_{22}$ are all memoryless operators with $\boldsymbol{T}_{11} > \boldsymbol{0}$ and $\boldsymbol{T}_{22} > 0$. From this it is an easy calculation to show that the operators

$$\begin{aligned}
\tilde{\boldsymbol{T}}_{11} &:= \Omega \boldsymbol{T}_{11}^* \Omega \\
\tilde{\boldsymbol{T}}_{12} &:= \Omega \boldsymbol{T}_{21}^* \Omega \\
\tilde{\boldsymbol{T}}_{22} &:= \Omega \boldsymbol{T}_{22}^* \Omega
\end{aligned}$$

satisfy the requirements. ∎

To find the minimum entropy closed-loop system, we first parameterize all closed-loop systems satisfying the norm bound. This characterization will be used to choose a controller which has attractive features regarding entropy.

Lemma 6.11 *All closed-loop systems $\mathcal{F}_\ell(\boldsymbol{G}_{FC}, \boldsymbol{K})$ satisfying*

$$\boldsymbol{I} - \mathcal{F}_\ell(\boldsymbol{G}_{FC}, \boldsymbol{K})^* \mathcal{F}_\ell(\boldsymbol{G}_{FC}, \boldsymbol{K}) > \boldsymbol{0}$$

are given by $\mathcal{F}_\ell(\Omega \boldsymbol{P}^* \Omega, \Omega \boldsymbol{Q}^* \Omega)$, where $\|\boldsymbol{Q}\| < 1$, and $\Omega \boldsymbol{P}^* \Omega$ is the operator given by the state-space realization

$$\left[\begin{array}{c|cc} \boldsymbol{A} + \boldsymbol{L}\begin{bmatrix} \tilde{\boldsymbol{T}}_{12}\tilde{\boldsymbol{T}}_{22}^{-1} \\ \boldsymbol{I} \end{bmatrix}\boldsymbol{C}_2 & \boldsymbol{B}_1 + \boldsymbol{L}\begin{bmatrix} \tilde{\boldsymbol{T}}_{12}\tilde{\boldsymbol{T}}_{22}^{-1} \\ \boldsymbol{I} \end{bmatrix}\boldsymbol{D}_{21} & -\boldsymbol{L}_1\tilde{\boldsymbol{T}}_{11} \\ \hline \boldsymbol{C}_1 - \tilde{\boldsymbol{T}}_{12}\tilde{\boldsymbol{T}}_{22}^{-1}\boldsymbol{C}_2 & \boldsymbol{D}_{11} - \tilde{\boldsymbol{T}}_{12}\tilde{\boldsymbol{T}}_{22}^{-1}\boldsymbol{D}_{21} & \tilde{\boldsymbol{T}}_{11} \\ \tilde{\boldsymbol{T}}_{22}^{-1}\boldsymbol{C}_2 & \tilde{\boldsymbol{T}}_{22}^{-1}\boldsymbol{D}_{21} & \boldsymbol{0} \end{array}\right] \tag{6.8}$$

where $\boldsymbol{L} = \begin{bmatrix} \boldsymbol{L}_1 & \boldsymbol{L}_2 \end{bmatrix}$ is defined by

$$\boldsymbol{L} := -(\boldsymbol{A}\boldsymbol{Z}^*\boldsymbol{Y}\boldsymbol{Z}\boldsymbol{C}^* + \boldsymbol{B}_1\boldsymbol{D}_{\bullet 1}^*)(\tilde{\boldsymbol{R}} + \boldsymbol{C}\boldsymbol{Z}^*\boldsymbol{Y}\boldsymbol{Z}\boldsymbol{C}^*)^{-1} \tag{6.9}$$

Proof: From Theorem 6.7 we know that all closed-loop systems satisfying the norm constraint $\|\mathcal{F}_\ell(\Omega \boldsymbol{G}_{FC}^* \Omega, \Omega \boldsymbol{K}^* \Omega)\| < 1$ are given by $\mathcal{F}_\ell(\boldsymbol{P}, \boldsymbol{Q})$, where $\boldsymbol{Q}$ is a causal operator with $\|\boldsymbol{Q}\| < 1$, and $\boldsymbol{P}$ equals the operator $\left[\begin{smallmatrix} \boldsymbol{P}_{11} & \boldsymbol{P}_{12} \\ \boldsymbol{P}_{21} & \boldsymbol{P}_{22} \end{smallmatrix}\right]$ given by the state-space realization

$$\left[\begin{array}{c|cc} \Omega \boldsymbol{A}^*\Omega + \Omega \boldsymbol{C}_2^*\Omega \boldsymbol{H} & \Omega \boldsymbol{C}_1^*\Omega - \Omega \boldsymbol{C}_2^*\Omega \boldsymbol{T}_{22}^{-1}\boldsymbol{T}_{21} & \Omega \boldsymbol{C}_2^*\Omega \boldsymbol{T}_{22}^{-1} \\ \hline \Omega \boldsymbol{B}_1^*\Omega + \Omega \boldsymbol{D}_{11}^*\Omega \boldsymbol{H} & \Omega \boldsymbol{D}_{11}^*\Omega - \Omega \boldsymbol{D}_{21}^*\Omega \boldsymbol{T}_{22}^{-1}\boldsymbol{T}_{21} & \Omega \boldsymbol{D}_{21}^*\Omega \boldsymbol{T}_{22}^{-1} \\ -\boldsymbol{T}_{11}\boldsymbol{F}_1 & \boldsymbol{T}_{11} & \boldsymbol{0} \end{array}\right]$$

Here

$$\begin{aligned} \boldsymbol{H} &:= \begin{bmatrix} \boldsymbol{T}_{22}^{-1}\boldsymbol{T}_{21} & \boldsymbol{I} \end{bmatrix}\boldsymbol{F} \\ \boldsymbol{F} &= \begin{bmatrix} \boldsymbol{F}_1 \\ \boldsymbol{F}_2 \end{bmatrix} := -\left(\boldsymbol{R} + \begin{bmatrix} \Omega & \boldsymbol{0} \\ \boldsymbol{0} & \Omega \end{bmatrix}\boldsymbol{C}\Omega \boldsymbol{Z}\boldsymbol{X}\boldsymbol{Z}^*\Omega \boldsymbol{C}^*\begin{bmatrix} \Omega & \boldsymbol{0} \\ \boldsymbol{0} & \Omega \end{bmatrix}\right)^{-1} \\ &\quad \times\left(\begin{bmatrix} \Omega & \boldsymbol{0} \\ \boldsymbol{0} & \Omega \end{bmatrix}\boldsymbol{C}\Omega \boldsymbol{Z}\boldsymbol{X}\boldsymbol{Z}^*\Omega \boldsymbol{A}^*\Omega + \begin{bmatrix} \Omega & \boldsymbol{0} \\ \boldsymbol{0} & \Omega \end{bmatrix}\boldsymbol{D}_{\bullet 1}\boldsymbol{B}_1^*\Omega\right) \end{aligned}$$

From Lemma 2.14 we know that $\mathcal{F}_\ell(\boldsymbol{G}_{FC}, \boldsymbol{K}) = \Omega \mathcal{F}_\ell(\Omega \boldsymbol{G}_{FC}^* \Omega, \Omega \boldsymbol{K}^* \Omega)^* \Omega$, hence all closed-loop systems $\mathcal{F}_\ell(\boldsymbol{G}_{FC}, \boldsymbol{K})$ satisfying the norm bound can be written as $\Omega \mathcal{F}_\ell(\boldsymbol{P}, \boldsymbol{Q})^* \Omega = \mathcal{F}_\ell(\Omega \boldsymbol{P}^* \Omega, \Omega \boldsymbol{Q}^* \Omega)$. The state-space realization of $\Omega \boldsymbol{P}^* \Omega$, given by (6.8), follows immediately by recalling that

$$\Sigma_{\Omega P^* \Omega} = \begin{bmatrix} \Omega \boldsymbol{P}_{11}^* \Omega & \Omega \boldsymbol{P}_{21}^* \Omega \\ \Omega \boldsymbol{P}_{12}^* \Omega & \Omega \boldsymbol{P}_{22}^* \Omega \end{bmatrix}$$

and $\boldsymbol{L} = \begin{bmatrix} \Omega \boldsymbol{F}_1^* \Omega & \Omega \boldsymbol{F}_2^* \Omega \end{bmatrix}$. ■

While the natural dual allows us to characterize the set of all closed-loop systems, we are not able to use the natural dual to find the minimum entropy controller. For the FI case, as in the proof of Theorem 6.4, we computed a spectral factorization of $\mathcal{F}_\ell(\boldsymbol{P},\boldsymbol{Q})$. For the FC problem we need a spectral factorization of

$$\begin{aligned}\boldsymbol{I} - \mathcal{F}_\ell(\Omega\boldsymbol{P}^*\Omega, \Omega\boldsymbol{Q}^*\Omega)^*\mathcal{F}_\ell(\Omega\boldsymbol{P}^*\Omega, \Omega\boldsymbol{Q}^*\Omega)& \\ = \boldsymbol{I} - \Omega\mathcal{F}_\ell(\boldsymbol{P}^*, \boldsymbol{Q}^*)^*\mathcal{F}_\ell(\boldsymbol{P}^*, \boldsymbol{Q}^*)\Omega& \\ = \Omega(\boldsymbol{I} - \mathcal{F}_\ell(\boldsymbol{P}, \boldsymbol{Q})\mathcal{F}_\ell(\boldsymbol{P}, \boldsymbol{Q})^*)\Omega&\end{aligned}$$

That is, we need a co-spectral factorization of $\mathcal{F}_\ell(\boldsymbol{P},\boldsymbol{Q})$ in order to find a spectral factorization of $\boldsymbol{I} - \mathcal{F}_\ell(\Omega\boldsymbol{P}^*\Omega, \Omega\boldsymbol{Q}^*\Omega)^*\mathcal{F}_\ell(\Omega\boldsymbol{P}^*\Omega, \Omega\boldsymbol{Q}^*\Omega)$. A possible way to overcome this is by extending the operator $\boldsymbol{P}$ as

$$\boldsymbol{P}_e := \begin{bmatrix} \boldsymbol{P}_{11} & \boldsymbol{P}_{12e} \\ \boldsymbol{P}_{21} & \boldsymbol{P}_{22e} \end{bmatrix} \tag{6.10}$$

such that both $\boldsymbol{P}_{12e}$ and $\boldsymbol{P}_{22e}$ are causal, and that the operator (6.10) is both inner and co-inner. This means that also the operator $\Omega\boldsymbol{P}_e^*\Omega$ is both inner and co-inner, while $\Omega\boldsymbol{P}^*\Omega$ can only be shown to be co-inner. That this can be done is shown in [78]. However, it is not difficult to check — even for an LTI system — that the operator $\boldsymbol{P}_{12e}$ will in general not have a causal inverse. This property is essential in evaluating the entropy of the closed-loop system in the proof of Theorem 6.4. Therefore, instead of minimizing the entropy, we will minimize the average entropy.

Lemma 6.12 *Among the controllers admitting a state-space realization, a closed-loop system with minimum average entropy is given by* $\Omega\boldsymbol{P}_{11}^*\Omega$.

Proof: We know from Lemma 5.17 that, if $\boldsymbol{Q}$ admits a state-space realization

$$\begin{aligned}\limsup_{k\to\infty}\sum_{j=-k}^{k}\boldsymbol{E}(\mathcal{F}_\ell(\Omega\boldsymbol{P}^*\Omega,\Omega\boldsymbol{Q}^*\Omega))_j &= \limsup_{k\to\infty}\sum_{j=-k}^{k}\boldsymbol{E}_a(\mathcal{F}_\ell(\Omega\boldsymbol{P}\Omega,\Omega\boldsymbol{Q}\Omega))_j \\ &= \limsup_{k\to\infty}\sum_{j=-k}^{k}\boldsymbol{E}(\mathcal{F}_\ell(\boldsymbol{P},\boldsymbol{Q}))_j\end{aligned}$$

where the second equality follows from Lemma 4.18. From Theorem 6.4 we know that $\boldsymbol{Q}_{\min} = \boldsymbol{0}$ minimizes the entropy $\boldsymbol{E}(\mathcal{F}_\ell(\boldsymbol{P},\boldsymbol{Q}))$, hence it minimizes the average entropy. The corresponding closed-loop system, minimizing the average entropy in the FC problem, is given by

$$\mathcal{F}_\ell(\Omega\boldsymbol{P}^*\Omega, \Omega\boldsymbol{Q}_{\min}^*\Omega) = \mathcal{F}_\ell(\Omega\boldsymbol{P}^*\Omega, \boldsymbol{0}) = \Omega\boldsymbol{P}_{11}^*\Omega$$

For completeness we mention that $\Omega \boldsymbol{P}_{11}^{*} \Omega$ can be written in terms of the parameters of the plant as the operator with state-space realization

$$\Sigma_{\Omega P_{11}^{*}\Omega} = \left[\begin{array}{c|c} \boldsymbol{A} + \boldsymbol{L}\begin{bmatrix} \tilde{\boldsymbol{T}}_{12}\tilde{\boldsymbol{T}}_{22}^{-1} \\ \boldsymbol{I} \end{bmatrix} \boldsymbol{C}_2 & \boldsymbol{B}_1 + \boldsymbol{L}\begin{bmatrix} \tilde{\boldsymbol{T}}_{12}\tilde{\boldsymbol{T}}_{22}^{-1} \\ \boldsymbol{I} \end{bmatrix} \boldsymbol{D}_{21} \\ \hline \boldsymbol{C}_1 - \tilde{\boldsymbol{T}}_{12}\tilde{\boldsymbol{T}}_{22}^{-1}\boldsymbol{C}_2 & \boldsymbol{D}_{11} - \tilde{\boldsymbol{T}}_{12}\tilde{\boldsymbol{T}}_{22}^{-1}\boldsymbol{D}_{21} \end{array}\right] \tag{6.11}$$

where $\boldsymbol{L}$ is given by (6.9). ■

In the next section we will consider the disturbance feedforward problem.

6.5 Disturbance feedforward

To solve the minimum entropy control problem in the disturbance feedforward (DF) case, we follow a similar procedure as in [53]. We will first characterize all closed-loop systems using a compensator which has a fixed component and a free component satisfying a norm constraint. We will then use this characterization to choose the free component such that the entropy is minimized. In this section we consider systems of the form (6.1), with the assumptions *A(1)–A(6)*. In the DF problem we make two additional assumptions. First of all, we assume that $\boldsymbol{A} - \boldsymbol{B}_1\boldsymbol{D}_{21}^{\dagger}\boldsymbol{C}_2$ is UES, which is a stronger assumption than *A(6)*.

Secondly, we assume that the disturbance w is composed of two parts. The first of these, w_1, does not affect the state or the measurement output; the second component w_2, although affecting both state and measurement, can be estimated from the latter exactly. To make notation easier, we introduce a permutation of the input $\boldsymbol{w}$.

$$\Pi \boldsymbol{w} := \begin{bmatrix} \boldsymbol{w}_1 \\ \boldsymbol{w}_2 \end{bmatrix}$$

were $\boldsymbol{w}_1$ (resp. $\boldsymbol{w}_2$) has as k^{th} element w_{1_k} (resp. w_{2_k}). It is easily seen that this permutation can be represented as

$$\Pi = \begin{bmatrix} \Pi_1 \\ \Pi_2 \end{bmatrix} = \left[\begin{array}{c} \text{diag}\{[\ I \quad 0\ ]\} \\ \hline \text{diag}\{[\ 0 \quad I\ ]\} \end{array}\right] \tag{6.12}$$

with appropriate dimensions, and that $\Pi^*\Pi = \Pi\Pi^* = \boldsymbol{I}$. With $\bar{\boldsymbol{w}} := \Pi \boldsymbol{w}$

the system $\boldsymbol{G}_{DF}$, given by (6.1), is equivalent to the system $\bar{\boldsymbol{G}}_{DF}$, given by

$$\Sigma_{\bar{G}_{DF}} := \begin{cases} Zx = Ax + B_1\Pi^*\bar{w} + B_2u \\ z = C_1x + D_{11}\Pi^*\bar{w} + D_{12}u \\ y = C_2x + D_{21}\Pi^*\bar{w} \end{cases}$$

The measurement output is not influenced by w_1. Since $D_{21}D_{21}^* > 0$, we can therefore assume without loss of generality that $D_{21}\Pi^* = \begin{bmatrix} 0 & I \end{bmatrix}$. Furthermore, we define $D_\perp$ with appropriate dimensions such that

$$\begin{bmatrix} D_{21}\Pi^* \\ D_\perp \end{bmatrix} = \begin{bmatrix} 0 & I \\ I & 0 \end{bmatrix}$$

The assumption that w_1 does not influence the state is now equivalent to $B_1\Pi^*D_\perp = 0$, resulting in the system

$$\Sigma_{\bar{G}_{DF}} = \begin{cases} Zx = Ax + \begin{bmatrix} 0 & B_{12} \end{bmatrix}\bar{w} + B_2u \\ z = C_1x + \begin{bmatrix} D_{111} & D_{112} \end{bmatrix}\bar{w} + D_{12}u \\ y = C_2x + \begin{bmatrix} 0 & I \end{bmatrix}\bar{w} \end{cases} \tag{6.13}$$

The structure of $\boldsymbol{G}_{DF}$ will simplify the equations involved in solving the minimum entropy control problem.

6.5.1 Characterizing all closed-loop systems

It will be apparent that the conditions for the existence of a stabilizing controller satisfying a closed-loop constraint are identical to the conditions for the FI case.

Theorem 6.13 *There exists a stabilizing controller K such that*

$$I - \mathcal{F}_\ell(G_{DF}, K)^*\mathcal{F}_\ell(G_{DF}, K) > 0$$

if and only if there exist X and T satisfying the conditions in Theorem 6.5. Furthermore, if these conditions are satisfied, all closed-loop systems which achieve the norm bound are characterized by $\mathcal{F}_\ell(P, Q)$, where P is given by (6.4), $Q = \begin{bmatrix} \bar{Q}_1 & \bar{Q}_2 \end{bmatrix}\Pi$ with

$$\bar{Q}_1 = T_{22}^{-*}D_{12}^*D_{111}\bar{T}_{111}^{-1}$$

and $\bar{Q}_2$ is any causal operator such that $\|Q\| < 1$. Here $\bar{T}_{111}$ is the positive definite memoryless solution of

$$\bar{T}_{111}^*\bar{T}_{111} = I - D_{111}^*D_{111} + D_{111}^*D_{12}(D_{12}^*D_{12} + B_2^*ZXZ^*B_2)^{-1}D_{12}^*D_{111} \tag{6.14}$$

Proof: Necessity is trivial, since if there exists a controller $\boldsymbol{K}$ that solves the DF problem, then the controller $\boldsymbol{K}[\ C_2 \quad D_{21}\ ]$ will solve the FI problem.

We will now characterize all stabilizing controllers satisfying the norm bound, which proves the sufficiency part. First note that, if $\boldsymbol{X}$ and $\boldsymbol{T}$ satisfy the conditions in Theorem 6.5 using the parameters of the system $\boldsymbol{G}_{DF}$, then $\boldsymbol{X}$ and $\bar{\boldsymbol{T}}$ satisfy these conditions using the parameters of the system $\bar{\boldsymbol{G}}_{DF}$, where

$$
\begin{aligned}
\bar{\boldsymbol{T}}_{11} &:= \Pi \boldsymbol{T}_{11} \Pi^* \\
\bar{\boldsymbol{T}}_{21} &:= \boldsymbol{T}_{21} \Pi^* \\
\bar{\boldsymbol{T}}_{22} &:= \boldsymbol{T}_{22}
\end{aligned}
$$

We know from Theorem 6.7 that all control signals $\boldsymbol{u}$ satisfying the closed-loop norm bound $\|\mathcal{F}_\ell(\boldsymbol{G}_{DF}, \boldsymbol{K})\| < 1$ are given by

$$
\begin{aligned}
\boldsymbol{u} &= \boldsymbol{T}_{22}^{-1}\boldsymbol{v} - \boldsymbol{T}_{22}^{-1}\boldsymbol{T}_{21}\boldsymbol{w} + \left[\ \boldsymbol{T}_{22}^{-1}\boldsymbol{T}_{21} \quad \boldsymbol{I}\ \right]\boldsymbol{F}\boldsymbol{x} \\
\boldsymbol{r} &= \boldsymbol{T}_{11}(\boldsymbol{w} - \boldsymbol{F}_1\boldsymbol{x}) \\
\boldsymbol{v} &= \boldsymbol{Q}\boldsymbol{r}
\end{aligned}
$$

where $\|\boldsymbol{Q}\| < 1$. These equations can be written in an equivalent form using the parameters corresponding to $\bar{\boldsymbol{G}}_{DF}$.

$$
\begin{aligned}
\boldsymbol{u} &= \bar{\boldsymbol{T}}_{22}^{-1}\boldsymbol{v} - \bar{\boldsymbol{T}}_{22}^{-1}\bar{\boldsymbol{T}}_{21}\bar{\boldsymbol{w}} + \left[\ \bar{\boldsymbol{T}}_{22}^{-1}\bar{\boldsymbol{T}}_{21} \quad \boldsymbol{I}\ \right]\bar{\boldsymbol{F}}\boldsymbol{x} \\
\bar{\boldsymbol{r}} &= \bar{\boldsymbol{T}}_{11}(\bar{\boldsymbol{w}} - \bar{\boldsymbol{F}}_1\boldsymbol{x}) \\
\boldsymbol{v} &= \bar{\boldsymbol{Q}}\bar{\boldsymbol{r}}
\end{aligned}
\tag{6.15}
$$

where $\bar{\boldsymbol{F}}_1 := \Pi\boldsymbol{F}_1$, $\bar{\boldsymbol{F}}_2 := \boldsymbol{F}_2$, $\bar{\boldsymbol{r}} := \Pi\boldsymbol{r}$, and $\bar{\boldsymbol{Q}} := \boldsymbol{Q}\Pi^*$. Note that we have $\|\bar{\boldsymbol{Q}}\| = \|\boldsymbol{Q}\|$. Since $\boldsymbol{w}_1$ does not affect $\boldsymbol{x}$ or $\boldsymbol{y}$, it follows that $\bar{\boldsymbol{Q}}$ must be restricted so that the mapping from $\boldsymbol{w}_1$ to $\boldsymbol{u}$ is identically zero, that is

$$
\bar{\boldsymbol{Q}}\bar{\boldsymbol{T}}_{11}\boldsymbol{D}_\perp^* - \bar{\boldsymbol{T}}_{21}\boldsymbol{D}_\perp^* = 0 \tag{6.16}
$$

Without loss of generality the memoryless operator $\boldsymbol{T}_{11}$ can be chosen such that each diagonal element is a 2×2 upper triangular block. By definition of $\bar{\boldsymbol{T}}_{11}$ this means we can partition

$$
\bar{\boldsymbol{T}}_{11} = \begin{bmatrix} \bar{\boldsymbol{T}}_{111} & \bar{\boldsymbol{T}}_{112} \\ \boldsymbol{0} & \bar{\boldsymbol{T}}_{113} \end{bmatrix}
$$

Recall that

$$T_{11}^* T_{11} = I + T_{21}^* T_{21} - D_{11}^* D_{11} - B_1^* Z X Z^* B_1$$

from which the equation (6.14) for $\bar{T}_{111}$ immediately follows. Since

$$\bar{T}_{21} D_{\perp} = \bar{T}_{22}^{-*} (D_{12}^* D_{11} \Pi^* + B_2^* Z X Z^* B_1 \Pi^*) D_{\perp}^* = \bar{T}_{22}^{-*} D_{12}^* D_{111}$$

equation (6.16) can be solved for $\bar{Q}_1$, where we have partitioned $\bar{Q}$ as $\bar{Q} = [\ \bar{Q}_1 \quad \bar{Q}_2\ ]$, and

$$\bar{Q}_1 = \bar{T}_{22}^{-*} D_{12}^* D_{111} \bar{T}_{111}^{-1}$$

It follows that all controller signals can be obtained as a feedback from w_2 and x by this particular structure of $\bar{Q}$. Substituting this in (6.15) gives

$$\begin{aligned} u &= \bar{T}_{22}^{-1} \bar{Q} \bar{T}_{11} (\bar{w} - \bar{F}_1 x) + \bar{T}_{22}^{-1} \bar{T}_{21} (\bar{F}_1 x - \bar{w}) + \bar{F}_2 x \\ &= (\bar{T}_{22}^* \bar{T}_{22})^{-1} \left[D_{12}^* D_{111} \bar{T}_{111}^{-1} \quad \bar{T}_{22}^* \bar{Q}_2 \right] \begin{bmatrix} \bar{T}_{111} & \bar{T}_{112} \\ 0 & \bar{T}_{113} \end{bmatrix} (\bar{w} - \bar{F}_1 x) + \bar{F}_2 x \\ &\quad + (\bar{T}_{22}^* \bar{T}_{22})^{-1} \left[\ D_{12}^* D_{111} \quad D_{12}^* D_{112} + B_2^* Z X Z^* B_{12}\ \right] (\bar{F}_1 x - \bar{w}) \\ &= (\bar{T}_{22}^* \bar{T}_{22})^{-1} H (\bar{F}_{12} x - w_2) + \bar{F}_2 x - \bar{T}_{22}^{-1} \bar{Q}_2 \bar{T}_{113} (\bar{F}_{12} x - w_2) \end{aligned}$$

where

$$H := D_{12}^* D_{112} + B_2^* Z X Z^* B_{12} - D_{12}^* D_{111} \bar{T}_{111}^{-1} \bar{T}_{112} \tag{6.17}$$

Thus, all control signals u can be expressed as feedback from x and w_2. Since these are not available, an observer can be designed to estimate them from the output y as follows:

$$\begin{aligned} Z \hat{x} &= A \hat{x} + B_{12} \hat{w}_2 + B_2 u \\ \hat{w}_2 &= -C_2 \hat{x} + y \\ u &= \bar{T}_{22}^{-1} \bar{T}_{22}^{-*} H (\bar{F}_{12} \hat{x} - \hat{w}_2) + \bar{F}_2 \hat{x} - \bar{T}_{22}^{-1} \bar{Q}_2 \bar{T}_{113} (\bar{F}_{12} \hat{x} - \hat{w}_2) \end{aligned} \tag{6.18}$$

It is easy to check that

$$Z(x - \hat{x}) = (A - B_{12} C_2)(x - \hat{x})$$

and thus, for any $k_0 \in \mathbb{Z}$, the initial condition $x_{k_0} = \hat{x}_{k_0} = 0$ ensures that $x_k = \hat{x}_k$ and $w_{2_k} = \hat{w}_{2_k}$ for all $k \geq k_0$. Since A_P is UES, stability follows from the assumption that $A - B_{12} C_2 = A - B_1 D_{21}^{\dagger} C_2$ is UES. ■

6.5.2 DF minimum entropy controller

The characterization in Theorem 6.13 enables us to find the minimum entropy controller. We can split up the entropy in three parts, namely the entropy of the FI case, the entropy of the fixed component $\bar{\boldsymbol{Q}}_1$, and the entropy of a system depending on the free component $\bar{\boldsymbol{Q}}_2$. Obviously, the optimization has to be done with respect to this last part only.

Lemma 6.14 *The entropy is minimized by the controller with state-space realization $\Sigma_{\bar{K}_{\min}}$ given by*

$$\begin{aligned} Z\hat{\boldsymbol{x}} &= \boldsymbol{A}\hat{\boldsymbol{x}} + \boldsymbol{B}_{12}\hat{\boldsymbol{w}}_2 + \boldsymbol{B}_2\boldsymbol{u} \\ \hat{\boldsymbol{w}}_2 &= -\boldsymbol{C}_2\hat{\boldsymbol{x}} + \boldsymbol{y} \\ \boldsymbol{u} &= \bar{\boldsymbol{T}}_{22}^{-1}\bar{\boldsymbol{T}}_{22}^{-*}\boldsymbol{H}(\bar{\boldsymbol{F}}_{12}\hat{\boldsymbol{x}} - \hat{\boldsymbol{w}}_2) + \bar{\boldsymbol{F}}_2\hat{\boldsymbol{x}} \end{aligned}$$

where $\boldsymbol{H}$ is given by (6.17). Moreover, the minimum value of the entropy is given by

$$\boldsymbol{E}(\mathcal{F}_\ell(\boldsymbol{G}_{DF}, \boldsymbol{K}_{\min})) = \boldsymbol{E}(\boldsymbol{P}_{11}) + \boldsymbol{E}(\bar{\boldsymbol{Q}}_1)$$

Proof: As in the proof of Theorem 6.4 we find that

$$\boldsymbol{E}(\mathcal{F}_\ell(\boldsymbol{G}_{DF}, \boldsymbol{K}_{\min})) = \boldsymbol{E}(\boldsymbol{P}_{11}) + \boldsymbol{E}(\boldsymbol{Q}) \tag{6.19}$$

The first term in (6.19) does not depend on $\boldsymbol{Q}$. Hence, for finding the minimum entropy controller, we only have to consider the entropy of $\boldsymbol{Q}$. To evaluate $\boldsymbol{E}(\boldsymbol{Q})$, we want to find a spectral factor of $\boldsymbol{I} - \boldsymbol{Q}^*\boldsymbol{Q}$. It can easily be checked that $\boldsymbol{I} - \boldsymbol{Q}^*\boldsymbol{Q} = \boldsymbol{L}^*\boldsymbol{L}$ for

$$\boldsymbol{L} := \Pi^* \begin{bmatrix} \boldsymbol{R}_1 & \boldsymbol{R}_2 \\ 0 & \boldsymbol{R}_3 \end{bmatrix} \Pi = \Pi_1^*\boldsymbol{R}_1\Pi_1 + \Pi_1^*\boldsymbol{R}_2\Pi_2 + \Pi_2^*\boldsymbol{R}_3\Pi_2$$

where

$$\begin{aligned} \boldsymbol{R}_1 &= (\boldsymbol{I} - \bar{\boldsymbol{Q}}_1^*\bar{\boldsymbol{Q}}_1)^{1/2} \in \mathcal{M} \\ \boldsymbol{R}_2 &= -(\boldsymbol{I} - \bar{\boldsymbol{Q}}_1^*\bar{\boldsymbol{Q}}_1)^{-1/2}\bar{\boldsymbol{Q}}_1^*\bar{\boldsymbol{Q}}_2 \in \mathcal{C} \end{aligned}$$

The operators $\boldsymbol{R}_i$ are well-defined since $\|\bar{\boldsymbol{Q}}\| < 1$. Furthermore $\boldsymbol{R}_3$ is a spectral factor satisfying

$$\begin{aligned} \boldsymbol{R}_3^*\boldsymbol{R}_3 &= \boldsymbol{I} - \bar{\boldsymbol{Q}}_2^*\bar{\boldsymbol{Q}}_2 - \bar{\boldsymbol{Q}}_2^*\bar{\boldsymbol{Q}}_1(\boldsymbol{I} - \bar{\boldsymbol{Q}}_1^*\bar{\boldsymbol{Q}}_1)^{-1}\bar{\boldsymbol{Q}}_1^*\bar{\boldsymbol{Q}}_2 \\ &= \boldsymbol{I} - \bar{\boldsymbol{Q}}_2^*(\boldsymbol{I} - \bar{\boldsymbol{Q}}_1\bar{\boldsymbol{Q}}_1^*)^{-1}\bar{\boldsymbol{Q}}_2 \end{aligned}$$

which exists also because of the norm constraint on $\bar{\boldsymbol{Q}}$. It follows that $\boldsymbol{L}$ is a spectral factor. For the entropy the memoryless part of this spectral factor is required. Since both Π_1 and Π_2 are memoryless, we can use the definitions of these operators to get

$$L_{k,k} = \begin{bmatrix} R_{1_k} & R_{2_{k,k}} \\ 0 & R_{3_{k,k}} \end{bmatrix}$$

Hence

$$\begin{aligned}
L_{k,k}^T L_{k,k} &= \begin{bmatrix} R_{1_k}^T & 0 \\ R_{2_{k,k}}^T & R_{3_{k,k}}^T \end{bmatrix} \begin{bmatrix} R_{1_k} & R_{2_{k,k}} \\ 0 & R_{3_{k,k}} \end{bmatrix} \\
&= \begin{bmatrix} R_{1_k}^T R_{1_k} & R_{1_k}^T R_{2_{k,k}} \\ R_{2_{k,k}}^T R_{1_k} & R_{2_{k,k}}^T R_{2_{k,k}} + R_{3_{k,k}}^T R_{3_{k,k}} \end{bmatrix} \\
&= \begin{bmatrix} I & 0 \\ S_k^T & I \end{bmatrix} \begin{bmatrix} R_{1_k}^T R_{1_k} & 0 \\ 0 & R_{3_{k,k}}^T R_{3_{k,k}} \end{bmatrix} \begin{bmatrix} I & S_k \\ 0 & I \end{bmatrix}
\end{aligned}$$

The matrices S_k are not of interest for our purpose; for completeness we mention that they are given by $S_k = R_{1_k}^{-1} R_{2_{k,k}}$. Now it is easy to compute the entropy of $\boldsymbol{Q}$.

$$\begin{aligned}
\boldsymbol{E}(\boldsymbol{Q}) &= -\operatorname{diag}\left\{\ln\det\left(L_{k,k}^T L_{k,k}\right)\right\} \\
&= -\operatorname{diag}\{\ln(\det(R_{1_k}^T R_{1_k})\det(R_{3_{k,k}}^T R_{3_{k,k}})\} \\
&= -\operatorname{diag}\left\{\ln\det\left(R_{1_k}^T R_{1_k}\right)\right\} - \operatorname{diag}\left\{\ln\det\left(R_{3_{k,k}}^T R_{3_{k,k}}\right)\right\} \\
&= \boldsymbol{E}(\bar{\boldsymbol{Q}}_1) + \boldsymbol{E}((\boldsymbol{I} - \bar{\boldsymbol{Q}}_1\bar{\boldsymbol{Q}}_1^*)^{-1/2}\bar{\boldsymbol{Q}}_2)
\end{aligned}$$

where the last step follows immediately from the definitions of $\boldsymbol{R}_1$ and $\boldsymbol{R}_3$. Hence the entropy of $\boldsymbol{Q}$ can be separated in two parts. The first part equals the entropy of $\bar{\boldsymbol{Q}}_1$, which is fixed. The second part is the entropy of a system depending on the free parameter $\bar{\boldsymbol{Q}}_2$. Theorem 6.4 tells us immediately that this entropy is minimized uniquely by $\bar{\boldsymbol{Q}}_2 = \boldsymbol{0}$. The optimal controller $\bar{\boldsymbol{K}}_{\min}$ follows from (6.18), and the minimum value of the entropy is given by (6.19) as

$$\boldsymbol{E}\big(\mathcal{F}_\ell(\boldsymbol{G}_{DF}, \boldsymbol{K}_{\min})\big) = \boldsymbol{E}\big(\mathcal{F}_\ell(\boldsymbol{P}, \boldsymbol{Q}_{\min})\big) = \boldsymbol{E}(\boldsymbol{P}_{11}) + \boldsymbol{E}(\bar{\boldsymbol{Q}}_1) \tag{6.20}$$

which completes the proof. ■

Remark 6.15 In the FI problem, the minimum value of the entropy equals the first term of the above expression (see (6.6)). Expression (6.20) shows that the increase in entropy, whenever full information is not available, is $\boldsymbol{E}(\bar{\boldsymbol{Q}}_1)$.

Remark 6.16 $\bar{\boldsymbol{K}}_{\min}$ is the minimum entropy controller for $\bar{\boldsymbol{G}}_{DF}$, which (see (6.13)) is just the original system $\boldsymbol{G}_{DF}$ in which the disturbance is permuted such that $\boldsymbol{w}_1$ and $\boldsymbol{w}_2$ are separated. Due to this separation the controller has a relatively straightforward form. With the definition of Π the controller $\boldsymbol{K}_{\min}$ for $\boldsymbol{G}_{DF}$ can be written using an estimator for $\Pi_2 \boldsymbol{w} = \boldsymbol{w}_2$, the components w_{2k} which are available for estimation.

In the next section we will consider the dual of the DF problem, the output estimation problem.

6.6 Output estimation

For the output estimation problem we consider systems of the form (6.1) satisfying assumptions *A(1)–A(6)*. For the output estimation (OE) problem we make two additional assumptions, which are the duals of the assumptions made in the DF problem. First of all, we assume that $\boldsymbol{A} - \boldsymbol{B}_2 \boldsymbol{D}_{12}^{\dagger} \boldsymbol{C}_1$ is UES, which is a stronger assumption than *A(5)*.

Secondly, we assume that the first component of the output at each time, say z_{k1}, is influenced by neither the state nor the control input, and the second component, say z_{k2}, can be controlled by the input completely. We can introduce the operator Π as in (6.12) with appropriate dimensions such that

$$\Pi \boldsymbol{z} := \begin{bmatrix} z_1 \\ z_2 \end{bmatrix}$$

Since $\boldsymbol{D}_{12}^* \boldsymbol{D}_{12} > \boldsymbol{0}$, we can assume without loss of generality that

$$\Pi \begin{bmatrix} \boldsymbol{C}_1 & \boldsymbol{D}_{11} & \boldsymbol{D}_{12} \end{bmatrix} = \begin{bmatrix} \boldsymbol{0} & \boldsymbol{D}_{111} & \boldsymbol{0} \\ \boldsymbol{C}_{12} & \boldsymbol{D}_{112} & \boldsymbol{I} \end{bmatrix} \tag{6.21}$$

The OE plant can be related to the DF plant, and the proof of the following theorem is very similar to the one for the FC case. Again we look at the dual $\Omega \boldsymbol{G}_{OE}^* \Omega$ of $\boldsymbol{G}_{OE}$, which has a state-space realization

$$\Sigma_{\Omega G_{OE}^* \Omega} = \begin{cases} \boldsymbol{Z}\boldsymbol{x} = \Omega \boldsymbol{A}^* \Omega \boldsymbol{x} + \Omega \boldsymbol{C}_1^* \Omega \boldsymbol{w} + \Omega \boldsymbol{C}_2^* \Omega \boldsymbol{u} \\ \boldsymbol{z} = \Omega \boldsymbol{B}_1^* \Omega \boldsymbol{x} + \Omega \boldsymbol{D}_{11}^* \Omega \boldsymbol{w} + \Omega \boldsymbol{D}_{21}^* \Omega \boldsymbol{u} \\ \boldsymbol{y} = \Omega \boldsymbol{B}_2^* \Omega \boldsymbol{x} + \Omega \boldsymbol{D}_{12}^* \Omega \boldsymbol{w} \end{cases}$$

Using the right dimensions, it is easily checked that

$$\Omega\Pi^* = \Pi^* \begin{bmatrix} \Omega & \mathbf{0} \\ \mathbf{0} & \Omega \end{bmatrix}$$

Hence the decomposition (6.21) can be written as

$$\begin{bmatrix} \Omega \boldsymbol{C}_1^* \Omega \\ \Omega \boldsymbol{D}_{11}^* \Omega \\ \Omega \boldsymbol{D}_{12}^* \Omega \end{bmatrix} \Pi^* = \begin{bmatrix} \mathbf{0} & \Omega \boldsymbol{C}_{12}^* \Omega \\ \Omega \boldsymbol{D}_{111}^* \Omega & \Omega \boldsymbol{D}_{112}^* \Omega \\ \mathbf{0} & \boldsymbol{I} \end{bmatrix}$$

Now we have a plant analogous to that of the DF problem. The following result follows immediately.

Theorem 6.17 *There exists a stabilizing controller $\boldsymbol{K}$ such that*

$$\boldsymbol{I} - \mathcal{F}_\ell(\boldsymbol{G}_{OE}, \boldsymbol{K})^* \mathcal{F}_\ell(\boldsymbol{G}_{OE}, \boldsymbol{K}) > 0$$

if and only if there exist $\boldsymbol{Y}$ and $\tilde{\boldsymbol{T}}$ satisfying the conditions in Theorem 6.10. Moreover, if these conditions are satisfied, among the controllers which admit a state-space realization, a closed-loop system with minimum average entropy is given by

$$\mathcal{F}_\ell(\boldsymbol{G}_{OE}, \boldsymbol{K}_{\min}) = \mathcal{F}_\ell(\Omega \boldsymbol{P}_{11}^* \Omega, \Omega\Pi^* \begin{bmatrix} \bar{\boldsymbol{Q}}_1 & \mathbf{0} \end{bmatrix}^* \Omega)$$

where $\Omega \boldsymbol{P}_{11}^* \Omega$ is the operator given in (6.11), and

$$\Omega \bar{\boldsymbol{Q}}_1^* \Omega = \hat{\boldsymbol{T}}_{111}^{-1} \boldsymbol{D}_{111} \boldsymbol{D}_{21}^* \tilde{\boldsymbol{T}}_{22}^{-1}$$

Here $\hat{\boldsymbol{T}}_{111}$ is a memoryless solution of

$$\hat{\boldsymbol{T}}_{111} \hat{\boldsymbol{T}}_{111}^* = \boldsymbol{I} - \boldsymbol{D}_{111} \boldsymbol{D}_{111}^* + \boldsymbol{D}_{111} \boldsymbol{D}_{21}^* (\boldsymbol{D}_{21} \boldsymbol{D}_{21}^* + \boldsymbol{C}_2 \boldsymbol{Z}^* \boldsymbol{Y} \boldsymbol{Z} \boldsymbol{C}_2^*)^{-1} \boldsymbol{D}_{21} \boldsymbol{D}_{111}^*$$

Proof: Completely analogous to the DF case, we can parameterize all closed-loop systems achieving the norm bound

$$\|\mathcal{F}_\ell(\Omega \boldsymbol{G}_{OE}^* \Omega, \Omega \boldsymbol{K}^* \Omega)\| < 1$$

as $\mathcal{F}_\ell(\boldsymbol{P}, \boldsymbol{Q})$, where $\boldsymbol{Q} = \begin{bmatrix} \bar{\boldsymbol{Q}}_1 & \bar{\boldsymbol{Q}}_2 \end{bmatrix} \Pi$ with

$$\bar{\boldsymbol{Q}}_1 = \boldsymbol{T}_{22}^{-1} \Omega \boldsymbol{D}_{21} \boldsymbol{D}_{111}^* \Omega \bar{\boldsymbol{T}}_{111}^{-1} \tag{6.22}$$

and $\bar{\boldsymbol{T}}_{111}$ is a memoryless solution of

$$\begin{aligned} \bar{\boldsymbol{T}}_{111}^* \bar{\boldsymbol{T}}_{111} &= \boldsymbol{I} - \Omega \boldsymbol{D}_{111} \boldsymbol{D}_{111}^* \Omega \\ &\quad + \Omega \boldsymbol{D}_{111} \boldsymbol{D}_{21}^* (\boldsymbol{D}_{21} \boldsymbol{D}_{21}^* + \boldsymbol{C}_2 \Omega \boldsymbol{Z} \boldsymbol{X} \boldsymbol{Z}^* \Omega \boldsymbol{C}_2^*)^{-1} \boldsymbol{D}_{21} \boldsymbol{D}_{111}^* \Omega \end{aligned}$$

Hence all closed-loop systems satisfying the norm bound $\|\mathcal{F}_\ell(\boldsymbol{G}_{OE}, \boldsymbol{K})\| < 1$ can be parameterized as $\mathcal{F}_\ell(\Omega \boldsymbol{P}^* \Omega, \Omega \boldsymbol{Q}^* \Omega)$.

Since $\bar{\boldsymbol{Q}}_2$ minimizes the entropy in the DF case, we can show, in exactly the same way as in the FC proof, that $\bar{\boldsymbol{Q}}_2 = \boldsymbol{0}$ minimizes the average entropy for the OE plant. The expression for $\bar{\boldsymbol{Q}}_1$ follows immediately from (6.22) by defining $\hat{\boldsymbol{T}}_{111} := \Omega \bar{\boldsymbol{T}}_{111}^* \Omega$. ■

The results for these special cases will enable us to solve the minimum average entropy control problem for the general output feedback case. This will be done in the next section.

6.7 Output feedback

In this section, the general output feedback problem is transformed to the output estimation special case of Section 6.6 through a change of variables. We consider systems of the form (6.1), and satisfying assumptions *A(1)–A(6)*.

Obviously, the FI problem has a solution if the general output feedback problem has one. Assuming the conditions in Theorem 6.5 are satisfied, we defined

$$\begin{aligned} \boldsymbol{v} &= \boldsymbol{T}_{22}\boldsymbol{u} + \boldsymbol{T}_{21}\boldsymbol{w} - \begin{bmatrix} \boldsymbol{T}_{21} & \boldsymbol{T}_{22} \end{bmatrix} \boldsymbol{F}\boldsymbol{x} \\ \boldsymbol{r} &= \boldsymbol{T}_{11}(\boldsymbol{w} - \boldsymbol{F}_1\boldsymbol{x}) \end{aligned}$$

By regarding $\boldsymbol{r}$ as input and $\boldsymbol{v}$ as output instead of $\boldsymbol{w}$ and $\boldsymbol{z}$ respectively, we get the operator $\boldsymbol{G}_{vyru}$ mapping $\left[\begin{smallmatrix} \boldsymbol{r} \\ \boldsymbol{u} \end{smallmatrix}\right]$ to $\left[\begin{smallmatrix} \boldsymbol{v} \\ \boldsymbol{y} \end{smallmatrix}\right]$ given by

$$\Sigma_{G_{vyru}} := \left[\begin{array}{c|cc} \boldsymbol{A} + \boldsymbol{B}_1\boldsymbol{F}_1 & \boldsymbol{B}_1\boldsymbol{T}_{11}^{-1} & \boldsymbol{B}_2 \\ \hline -\boldsymbol{T}_{22}\boldsymbol{F}_2 & \boldsymbol{T}_{21}\boldsymbol{T}_{11}^{-1} & \boldsymbol{T}_{22} \\ \boldsymbol{C}_2 + \boldsymbol{D}_{21}\boldsymbol{F}_1 & \boldsymbol{D}_{21}\boldsymbol{T}_{11}^{-1} & \boldsymbol{0} \end{array}\right]$$

Similarly, the operator $\boldsymbol{G}_{zrwv}$ is given by $\boldsymbol{P}$ as defined in (6.4). Using the state-space equations it is straightforward to see that $\mathcal{F}_\ell(\boldsymbol{G}, \boldsymbol{K})$ and $\mathcal{F}_\ell(\boldsymbol{P}, \mathcal{F}_\ell(\boldsymbol{G}_{vyru}, \boldsymbol{K}))$ are equivalent, as is shown in Figure 6.2.

Since $\boldsymbol{P}$ is an inner operator it follows immediately from Lemma 6.3 that $\boldsymbol{K}$ stabilizes $\boldsymbol{G}$ and $\|\mathcal{F}_\ell(\boldsymbol{G}, \boldsymbol{K})\| < 1$ if and only if $\|\mathcal{F}_\ell(\boldsymbol{G}_{vyru}, \boldsymbol{K})\| < 1$. For stability we need the system $\boldsymbol{G}_{vyru}$ to be stabilizable and detectable. The system is stabilizable since

$$\boldsymbol{A} + \boldsymbol{B}_1\boldsymbol{F}_1 + \boldsymbol{B}_2\boldsymbol{F}_2 = \boldsymbol{A}_{cl}$$

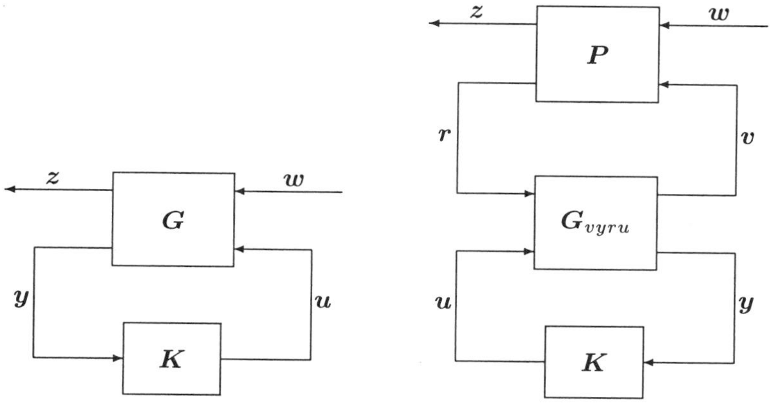

Figure 6.2: $\mathcal{F}_\ell(\boldsymbol{G}, \boldsymbol{K})$ and $\mathcal{F}_\ell(\boldsymbol{P}, \mathcal{F}_\ell(\boldsymbol{G}_{vyru}, \boldsymbol{K}))$

which is UES. The detectability condition is not straightforward, and we make the extra assumption

A(7) The pair $(\boldsymbol{C}_2 + \boldsymbol{D}_{21}\boldsymbol{F}_1, \boldsymbol{A} + \boldsymbol{B}_1\boldsymbol{F}_1)$ is uniformly detectable

This assumption is weaker than the one made in [67] where the corresponding continous-time, time-varying control problem is considered. It is, however, stronger than the conditions that are known for the time-invariant $\mathcal{H}_\infty$ control problem.

Remark 6.18 In the LTI case, assumption *A(6)* is replaced by the assumption that

$$\begin{bmatrix} A - zI & B_1 \\ C_2 & D_{21} \end{bmatrix} \tag{6.23}$$

has full row rank for all z on the unit circle. It can be shown that this condition implies *A(7)*. At this point it is not clear how (6.23) translates to the time-varying case. This will be discussed in more detail in Section 6.8.

Now, to show the connection between the general output feedback plant and a related OE plant, we take any operator $\boldsymbol{W} \in \mathcal{M}$ satisfying

$$\boldsymbol{W}^*\boldsymbol{W} = \boldsymbol{I} - \boldsymbol{T}_{11}^*\boldsymbol{T}_{11}$$

where the right-hand side is positive semi-definite since $\boldsymbol{P}$ is inner. As in the OE problem, we define the operator Π as in (6.21), and also the operator $\tilde{\boldsymbol{D}}_\perp$ of appropriate dimensions such that

$$\begin{bmatrix} \Omega \boldsymbol{D}_{12}^* \Omega \Pi^* \\ \Omega \tilde{\boldsymbol{D}}_\perp^* \Omega \Pi^* \end{bmatrix} = \begin{bmatrix} \mathbf{0} & \boldsymbol{I} \\ \boldsymbol{I} & \mathbf{0} \end{bmatrix}$$

Using this notation we can define the system $\boldsymbol{G}_{\text{tmp}}$, given by

$$\Sigma_{G_{\text{tmp}}} = \left[\begin{array}{c|c} \boldsymbol{A}_{\text{tmp}} & \boldsymbol{B}_{\text{tmp}} \\ \hline \boldsymbol{C}_{\text{tmp}} & \boldsymbol{D}_{\text{tmp}} \end{array}\right] := \left[\begin{array}{c|cc} \boldsymbol{A} + \boldsymbol{B}_1 \boldsymbol{F}_1 & \boldsymbol{B}_1 & \boldsymbol{B}_2 \boldsymbol{T}_{22}^{-1} \\ \hline -\boldsymbol{D}_{12} \boldsymbol{T}_{22} \boldsymbol{F}_2 & \boldsymbol{S} & \boldsymbol{D}_{12} \\ \boldsymbol{C}_2 + \boldsymbol{D}_{21} \boldsymbol{F}_1 & \boldsymbol{D}_{21} & \mathbf{0} \end{array}\right]$$

where $\boldsymbol{S} := \tilde{\boldsymbol{D}}_\perp \boldsymbol{W} + \boldsymbol{D}_{12} \boldsymbol{T}_{21}$.

Remark 6.19 According to the dimensions of the operators the multiplication $\tilde{\boldsymbol{D}}_\perp \boldsymbol{W}$ need not be well-defined. If the dimensions do not match, we can augment the output equation as

$$\boldsymbol{v} = \begin{bmatrix} \mathbf{0} \\ -\boldsymbol{D}_{12} \boldsymbol{T}_{22} \boldsymbol{F}_2 \end{bmatrix} \boldsymbol{x} + \left(\begin{bmatrix} \mathbf{0} \\ \boldsymbol{D}_{12} \end{bmatrix} \boldsymbol{T}_{21} + \begin{bmatrix} \mathbf{0} & \boldsymbol{I} \\ \tilde{\boldsymbol{D}}_\perp & \mathbf{0} \end{bmatrix} \boldsymbol{W} \right) \boldsymbol{w} + \begin{bmatrix} \mathbf{0} \\ \boldsymbol{D}_{12} \end{bmatrix} \boldsymbol{u}$$

where the identity operator is of appropriate dimensions.

The importance of the plant $\boldsymbol{G}_{\text{tmp}}$ is that it satisfies the requirements of the OE problem. This enables us to solve the problem of minimizing the entropy for the general output feedback problem. First we will give necessary and sufficient conditions for the existing of a stabilizing controller achieving a norm bound, which are obvious from the derivation above.

Theorem 6.20 *There exists a stabilizing controller $\boldsymbol{K}$ such that*

$$\boldsymbol{I} - \mathcal{F}_\ell(\boldsymbol{G}, \boldsymbol{K})^* \mathcal{F}_\ell(\boldsymbol{G}, \boldsymbol{K}) > 0$$

if and only if there exist $\boldsymbol{X}$ and $\boldsymbol{T}$ satisfying the conditions in Theorem 6.5 for the plant $\boldsymbol{G}$, and $\boldsymbol{Y}_{\text{tmp}}$ and $\tilde{\boldsymbol{T}}_{\text{tmp}}$ satisfying the conditions in Theorem 6.13 for the plant $\boldsymbol{G}_{\text{tmp}}$.

Proof: The conditions for the plant $\boldsymbol{G}$ are immediate, since the FI problem has a solution if the general problem has one. We just discussed that in that case we have to consider the interconnection $\mathcal{F}_\ell(\boldsymbol{G}_{vyru}, \boldsymbol{K})$. Since

$$\mathcal{F}_\ell(\boldsymbol{G}_{\text{tmp}}, \boldsymbol{T}_{22}\boldsymbol{K}) = \begin{bmatrix} \tilde{\boldsymbol{D}}_\perp & \boldsymbol{D}_{12} \end{bmatrix} \begin{bmatrix} \boldsymbol{W} \\ \mathcal{F}_\ell(\boldsymbol{G}_{vyru}, \boldsymbol{K}) \boldsymbol{T}_{11} \end{bmatrix}$$

it is a straightforward calculation to get

$$\begin{aligned} &\boldsymbol{I} - \mathcal{F}_\ell(\boldsymbol{G}_{\mathrm{tmp}}, \boldsymbol{T}_{22}\boldsymbol{K})^* \mathcal{F}_\ell(\boldsymbol{G}_{\mathrm{tmp}}, \boldsymbol{T}_{22}\boldsymbol{K}) \\ &\quad = \boldsymbol{T}_{11}^* \big(\boldsymbol{I} - \mathcal{F}_\ell(\boldsymbol{G}_{vyru}, \boldsymbol{K})^* \mathcal{F}_\ell(\boldsymbol{G}_{vyru}, \boldsymbol{K})\big)\boldsymbol{T}_{11} \end{aligned} \tag{6.24}$$

which shows that $\|\mathcal{F}_\ell(\boldsymbol{G}_{\mathrm{tmp}}, \boldsymbol{T}_{22}\boldsymbol{K})\| < 1$ if and only if $\|\mathcal{F}_\ell(\boldsymbol{G}_{vyru}, \boldsymbol{K})\| < 1$. For the system $\boldsymbol{G}_{\mathrm{tmp}}$ we have to check the assumptions made in the OE problem. *A(1)* and *A(2)* are obviously satisfied. *A(3)* is satisfied since

$$\boldsymbol{A}_{\mathrm{tmp}} + \boldsymbol{B}_{\mathrm{tmp}_2}\boldsymbol{T}_{22}\boldsymbol{F}_2 = \boldsymbol{A}_{cl}$$

which is UES. *A(4)* is satisfied under assumption *A(7)*, and since

$$\boldsymbol{A}_{\mathrm{tmp}} - \boldsymbol{B}_{\mathrm{tmp}_2}\boldsymbol{D}_{\mathrm{tmp}_{12}}^{\dagger}\boldsymbol{C}_{\mathrm{tmp}_1} = \boldsymbol{A}_{cl}$$

also the extra assumption in the OE problem, replacing *A(5)*, is satisfied. By applying a preliminary feedback $-\boldsymbol{F}_1$ assumption *A(6)* for $\boldsymbol{G}_{\mathrm{tmp}}$ is satisfied because of the same assumption for the original plant $\boldsymbol{G}$. So $\boldsymbol{G}_{\mathrm{tmp}}$ is an OE plant, and the conditions for $\boldsymbol{G}_{\mathrm{tmp}}$ follow immediately from the OE results. Since $\boldsymbol{T}_{22}$ is invertible, finding a controller $\boldsymbol{T}_{22}\boldsymbol{K}$ is equivalent to finding $\boldsymbol{K}$. ■

Remark 6.21 Along the lines of the proof in [52] it can be shown that the Riccati operator equations can be decoupled, as in the time-invariant case. This results in the statement that there exists a stabilizing controller $\boldsymbol{K}$ such that

$$\boldsymbol{I} - \mathcal{F}_\ell(\boldsymbol{G}, \boldsymbol{K})^* \mathcal{F}_\ell(\boldsymbol{G}, \boldsymbol{K}) > 0$$

if and only if there exist $\boldsymbol{X}$, $\boldsymbol{T}$, $\boldsymbol{Y}$ and $\tilde{\boldsymbol{T}}$ satisfying the conditions in Theorems 6.5 and 6.13 with $\rho(\boldsymbol{Z}\boldsymbol{X}\boldsymbol{Z}^*\boldsymbol{Y}) < 1$.

In order not to complicate the formulas we will express the minimum entropy in terms of $\boldsymbol{G}_{\mathrm{tmp}}$ rather than $\boldsymbol{G}$.

Lemma 6.22 *Among the controllers which admit a state-space realization, a closed-loop system with minimum average entropy is given by*

$$\mathcal{F}_\ell(\boldsymbol{G}, \boldsymbol{K}_{\min}) = \mathcal{F}_\ell(\boldsymbol{G}_{\mathrm{tmp}}, \boldsymbol{T}_{22}\boldsymbol{K}_{\min})$$

where $\boldsymbol{T}_{22}\boldsymbol{K}_{\min}$ *is the controller minimizing the average entropy of the OE plant* $\boldsymbol{G}_{\mathrm{tmp}}$.

Proof: Regarding $\mathcal{F}_\ell(\boldsymbol{G}_{vyru}, \boldsymbol{K})$ as $\boldsymbol{Q}$ in the proof of Theorem 6.4 we get

$$\begin{aligned}\boldsymbol{E}(\mathcal{F}_\ell(\boldsymbol{G},\boldsymbol{K})) &= \boldsymbol{E}(\mathcal{F}_\ell(\boldsymbol{P},\mathcal{F}_\ell(\boldsymbol{G}_{vyru},\boldsymbol{K}))) \\ &= \boldsymbol{E}(\boldsymbol{P}_{11}) + \boldsymbol{E}(\mathcal{F}_\ell(\boldsymbol{G}_{vyru},\boldsymbol{K}))\end{aligned}$$

From (6.24) we know that

$$\begin{aligned}\boldsymbol{E}\big(\mathcal{F}_\ell(\boldsymbol{G}_{vyru},\boldsymbol{K})\big) &= \boldsymbol{E}\big(\mathcal{F}_\ell(\boldsymbol{G}_{\text{tmp}},\boldsymbol{T}_{22}\boldsymbol{K})\big) + \text{diag}\left\{\ln\det\left(T_{11_k}^T T_{11_k}\right)\right\} \\ &= \boldsymbol{E}\big(\mathcal{F}_\ell(\boldsymbol{G}_{\text{tmp}},\boldsymbol{T}_{22}\boldsymbol{K})\big) - \boldsymbol{E}(\boldsymbol{P}_{11})\end{aligned}$$

The result follows from the fact that $\boldsymbol{G}_{\text{tmp}}$ is an OE plant. ∎

Remark 6.23 It can be shown that the system

$$\boldsymbol{G}_{vyru}\begin{bmatrix}\boldsymbol{I} & 0 \\ 0 & \boldsymbol{T}_{22}^{-1}\end{bmatrix}$$

also satisfies the condition for the OE problem. Hence the entropy corresponding to a closed-loop system with minimum average entropy can also be written as

$$\boldsymbol{E}\big(\mathcal{F}_\ell(\boldsymbol{G},\boldsymbol{K}_{\min})\big) = \boldsymbol{E}(\boldsymbol{P}_{11}) + \boldsymbol{E}\left(\mathcal{F}_\ell\left(\boldsymbol{G}_{vyru}\begin{bmatrix}\boldsymbol{I} & 0 \\ 0 & \boldsymbol{T}_{22}^{-1}\end{bmatrix}, \boldsymbol{T}_{22}\boldsymbol{K}_{\min}\right)\right)$$

Notice that the first term is the entropy corresponding to the FI case, and the second term corresponds to a related OE plant. This separation principle is similar to the time-invariant case.

Remark 6.24 For all the problems considered in this chapter, a state-space realization for a minimum average entropy controller can be obtained directly by applying the results from the DF case to its dual, the OE problem. Lemma 6.22 has the nice structure that the $\boldsymbol{D}_{12}$ and $\boldsymbol{D}_{21}$ operators of $\boldsymbol{G}_{\text{tmp}}$ are the same as those in the original plant. On the other hand, the operator

$$\boldsymbol{G}_{vyru}\begin{bmatrix}\boldsymbol{I} & 0 \\ 0 & \boldsymbol{T}_{22}^{-1}\end{bmatrix}$$

has the attractive feature that its $\boldsymbol{D}_{12}$ component equals the identity operator. Therefore it is a special OE plant, in the sense that the fixed component $\bar{\boldsymbol{Q}}_1$ in the OE solution is empty, which makes the formulae easier.

6.8 Stability concepts

In Remark 6.18 we noted one of the difficulties that arises in translating some of the assumptions from the LTI case to the time-varying problem. In this section we present a conjecture, which we believe could lead to an interesting theoretical framework for dealing with these difficulties.

For linear time-invariant systems the concepts of reachability, stabilizability, detectability and observability can be characterized in terms of eigenvalue tests, known as the Popov-Belevitch-Hautus test; see for example [70]. For example the pair (C, A) is observable if and only if

$$\left.\begin{array}{rcl} Ax &=& \lambda x \\ Cx &=& 0 \end{array}\right\} \Longrightarrow x = 0 \tag{6.25}$$

Tests like this are very attractive from a theoretical point of view, and are used repeatedly in proving these properties for systems. For time-varying systems however, eigenvalues need not exist. In this section we will give a conjecture stating an equivalent test for the systems we considered in this book. We will discuss the observability test, other tests can be derived similarly.

Consider the concept of uniform observability for discrete-time time-varying systems [70]: the bounded sequences $\{C_k\}$ and $\{A_k\}$ are uniformly observable if there exist a finite $l \in \mathbb{N}$ and an $\epsilon > 0$ such that

$$\sum_{r=k}^{k+l-l} A_k^T \cdots A_{r-1}^T C_r^T C_r A_{r-1} \cdots A_k > \epsilon I \tag{6.26}$$

for all $k \in \mathbb{Z}$. Using our infinite-dimensional operators $\boldsymbol{A} = \text{diag}\{A_k\}$ and $\boldsymbol{C} = \text{diag}\{C_k\}$, it is not difficult to write this as

$$\boldsymbol{M}_l := \sum_{j=0}^{l-1} (\boldsymbol{A}^* \boldsymbol{Z})^j \boldsymbol{C}^* \boldsymbol{C} (\boldsymbol{Z}^* \boldsymbol{A})^j > \epsilon \boldsymbol{I} \tag{6.27}$$

It can be shown that, if $\boldsymbol{M}_l > \epsilon \boldsymbol{I}$, the operator

$$\boldsymbol{H} := -(\boldsymbol{A}\boldsymbol{Z}^*)^{l+1} \boldsymbol{M}_{l+1}^{-1} (\boldsymbol{Z}\boldsymbol{A}^*)^l \boldsymbol{Z}\boldsymbol{C}^*$$

is such that $\boldsymbol{A} + \boldsymbol{H}\boldsymbol{C}$ is UES.

We are interested in finding a necessary and sufficient condition for uniform observability, analogous to the condition (6.25) for LTI systems. Since an infinite-dimensional operator need not have eigenvalues, we will consider the non-empty set of *almost eigenvalues* [14]. A number $\lambda \in \mathbb{C}$ is an almost eigenvalue of an operator $\boldsymbol{T}$ if there exists a sequence $\{\boldsymbol{x}_n\}_{n\geq 0}$ with $\|\boldsymbol{x}_n\|_2 = 1$ such that

$$\|\boldsymbol{T}\boldsymbol{x}_n - \lambda\boldsymbol{x}_n\|_2 \to 0 \quad (n \to \infty)$$

The sequence $\{\boldsymbol{x}_n\}$ is called an *almost eigenvector* corresponding to λ.

Theorem 6.25 *If the pair $(\boldsymbol{C}, \boldsymbol{A})$ is uniformly observable, then for every $\lambda \in \mathbb{C}$*

$$\left.\begin{array}{rcl} \|\boldsymbol{Z}^*\boldsymbol{A}\boldsymbol{x}_n - \lambda\boldsymbol{x}_n\|_2 & \to & 0 \\ \|\boldsymbol{C}\boldsymbol{x}_n\|_2 & \to & 0 \end{array}\right\} \Longrightarrow \|\boldsymbol{x}_n\|_2 \to 0$$

Proof: If λ is an almost eigenvalue of $\boldsymbol{Z}^*\boldsymbol{A}$ with corresponding almost eigenvector $\boldsymbol{x}_n$, we can assume that $\boldsymbol{x}_n$ is such that

$$\boldsymbol{Z}^*\boldsymbol{A}\boldsymbol{x}_n = \lambda\boldsymbol{x}_n + \boldsymbol{q}_n$$

where $\|\boldsymbol{q}_n\|_2 \leq \frac{1}{n}$. By assumption $\|\boldsymbol{C}\boldsymbol{x}_n\|_2 \to 0$, so it can assumed that $\|\boldsymbol{C}\boldsymbol{x}_n\|_2 \leq \frac{1}{n}$. For $j \geq 1$ we find that

$$\begin{aligned} \|\boldsymbol{C}(\boldsymbol{Z}^*\boldsymbol{A})^j\boldsymbol{x}_n\|_2 &= \|\boldsymbol{C}(\boldsymbol{Z}^*\boldsymbol{A})^{j-1}(\lambda\boldsymbol{x}_n + \boldsymbol{q}_n)\|_2 \\ &\leq |\lambda|\,\|\boldsymbol{C}(\boldsymbol{Z}^*\boldsymbol{A})^{j-1}\boldsymbol{x}_n\|_2 + \|\boldsymbol{C}(\boldsymbol{Z}^*\boldsymbol{A})^{j-1}\|\,\|\boldsymbol{q}_n\|_2 \\ &\leq |\lambda|\,\|\boldsymbol{C}(\boldsymbol{Z}^*\boldsymbol{A})^{j-1}\boldsymbol{x}_n\|_2 + \|\boldsymbol{C}\|\,\|\boldsymbol{Z}^*\boldsymbol{A}\|^{j-1}\,\frac{1}{n} \end{aligned}$$

Using recursion, it follows that for all $j \geq 1$

$$\|\boldsymbol{C}(\boldsymbol{Z}^*\boldsymbol{A})^j\boldsymbol{x}_n\|_2 \leq \frac{|\lambda|^j}{n} + \sum_{r=0}^{j-1}\|\boldsymbol{C}\|\,\|\boldsymbol{Z}^*\boldsymbol{A}\|^{j-r-1}\,\frac{|\lambda|^r}{n}$$

Now, by defining $\kappa := \max\{1, \|(\boldsymbol{Z}^*\boldsymbol{A})\|^{l-1}\}$, we see that for $1 \leq j \leq l-1$ this is bounded above by

$$\|\boldsymbol{C}(\boldsymbol{Z}^*\boldsymbol{A})^j\boldsymbol{x}_n\|_2 \leq \frac{|\lambda|^j}{n} + \sum_{r=0}^{j-1}\|\boldsymbol{C}\|\,\kappa\,\frac{|\lambda|^r}{n}$$

By applying this to the left-hand side of (6.27) we see that

$$\begin{aligned} \sum_{j=0}^{l-1}\|\boldsymbol{C}(\boldsymbol{Z}^*\boldsymbol{A})^j\boldsymbol{x}_n\|_2 &\leq \sum_{j=0}^{l-1}\frac{|\lambda|^j}{n} + \sum_{j=1}^{l-1}\sum_{r=0}^{j-1}\|\boldsymbol{C}\|\,\kappa\,\frac{|\lambda|^r}{n} \\ &= \frac{1-|\lambda|^l}{1-|\lambda|}\frac{1}{n} + \|\boldsymbol{C}\|\,\kappa\,\frac{1-2|\lambda|+|\lambda|^l}{(1-|\lambda|)^2}\,\frac{1}{n} \end{aligned}$$

whenever $|\lambda| \neq 1$. For $|\lambda| = 1$ it is easy to compute

$$\sum_{j=0}^{l-1} \|C(Z^*A)^j x_n\|_2 \leq l\,\frac{1}{n} + \|C\|\,\kappa\,\frac{l(l+1)}{2}\frac{1}{n}$$

It follows that

$$\langle x_n, M_l x_n\rangle = \sum_{j=0}^{l-1} \|C(Z^*A)^j x_n\|_2^2 \to 0 \quad (n \to \infty)$$

Since this is an upper bound for the right-hand side of (6.27), we can conclude that

$$\epsilon\|x_n\|_2^2 < \langle x_n, M_l x_n\rangle \to 0 \quad (n \to \infty)$$

Since $\epsilon > 0$ is fixed, the result follows. ■

Hence we have shown that, if the pair (C, A) is uniformly observable, there does not exist an almost eigenvalue of Z^*A for which the corresponding almost eigenvector x_n is such that $\|Cx_n\|_2$ goes to zero. Our belief is that the converse is also true.

Conjecture 6.26 *If for every $\lambda \in \mathbb{C}$*

$$\left.\begin{array}{rcl} \|Z^*Ax_n - \lambda x_n\|_2 & \to & 0 \\ \|Cx_n\|_2 & \to & 0 \end{array}\right\} \Longrightarrow \|x_n\|_2 \to 0$$

Then the pair (C, A) is uniformly observable. □

While we have not been able to prove this result, we believe that the following result might be useful. It shows that in considering the set of almost eigenvalues, we need only consider real ones.

Lemma 6.27 *If, for $A = \operatorname{diag}\{A_k\}$, λ is an almost eigenvalue of Z^*A, then $|\lambda|$ is also an almost eigenvalue of Z^*A, and it has a corresponding real almost eigenvector.*

Proof: Say $\|Z^*Ax_n - \lambda x_n\|_2 \to 0$ as $n \to \infty$, for $\|x_n\|_2 = 1$. Hence, with $\lambda = |\lambda|e^{i\theta}$, we have

$$\sum_{k=-\infty}^{\infty} \left|A_k x_{n_k} - |\lambda|e^{i\theta} x_{n_{k+1}}\right|^2 \to 0 \quad (n \to \infty)$$

This is equivalent to

$$\sum_{k=-\infty}^{\infty} \left| A_k e^{ik\theta} x_{n_k} - |\lambda| e^{i(k+1)\theta} x_{n_{k+1}} \right|^2 \to 0 \quad (n \to \infty) \tag{6.28}$$

which shows that $\tilde{\boldsymbol{x}}_n$, given by $\tilde{x}_{n_k} = e^{ik\theta} x_{n_k}$, is an almost eigenvector for the almost eigenvalue $|\lambda|$. Note that $\|\tilde{\boldsymbol{x}}_n\|_2 = \|\boldsymbol{x}_n\|_2 = 1$. To construct a real almost eigenvector, we can separate the real and the imaginary part in (6.28), which gives

$$\sum_{k=-\infty}^{\infty} \left| A_k Re(\tilde{x}_{n_k}) - |\lambda| Re(\tilde{x}_{n_{k+1}}) \right|^2 + \left| A_k Im(\tilde{x}_{n_k}) - |\lambda| Im(\tilde{x}_{n_{k+1}}) \right|^2 \to 0$$

as n goes to infinity. Hence we can take either $\boldsymbol{y}_{1n} := \dfrac{Re(\tilde{\boldsymbol{x}}_n)}{\|Re(\tilde{\boldsymbol{x}}_n)\|_2}$ or $\boldsymbol{y}_{2n} := \dfrac{Im(\tilde{\boldsymbol{x}}_n)}{\|Im(\tilde{\boldsymbol{x}}_n)\|_2}$ in order to have a real almost eigenvector. Note that at least one of them is well-defined. ∎

Remark 6.28 In a similar way we can show that $|\lambda| e^{i\psi}$ is an almost eigenvalue for every $\psi \in [0, 2\pi)$. This generalizes the well-known result that for the shift operator $\boldsymbol{Z}^*$ (i.e. $\boldsymbol{A} = \boldsymbol{I}$), the whole unit circle is part of the spectrum.

Remark 6.29 As an example of Lemma 6.27 we consider the linear time-invariant system with

$$A_k = A = \begin{bmatrix} 0 & 1 \\ -1 & 0 \end{bmatrix}$$

In this case $\lambda = i$ is an almost eigenvalue of $\boldsymbol{Z}^* \boldsymbol{A}$ with almost eigenvector $\boldsymbol{x}_n$ given by

$$x_{n_k} = \begin{cases} \dfrac{1}{\sqrt{2(2n+1)}} \begin{bmatrix} 1 \\ i \end{bmatrix} & \text{for } k = -n, \ldots, n-1, n \\ \begin{bmatrix} 0 \\ 0 \end{bmatrix} & \text{for } k \neq -n, \ldots, n-1, n \end{cases}$$

Following the construction in the lemma we find a real almost eigenvector $\boldsymbol{y}_n$, corresponding to $|\lambda| = 1$, given by

$$y_{n_0} = \frac{1}{\sqrt{(2n+1)}} \begin{bmatrix} 1 \\ 0 \end{bmatrix} \qquad y_{n_1} = \frac{1}{\sqrt{(2n+1)}} \begin{bmatrix} 0 \\ -1 \end{bmatrix}$$

$$y_{n_2} = \frac{1}{\sqrt{(2n+1)}} \begin{bmatrix} -1 \\ 0 \end{bmatrix} \qquad y_{n_3} = \frac{1}{\sqrt{(2n+1)}} \begin{bmatrix} 0 \\ 1 \end{bmatrix}$$

$$y_{n_k} = \begin{cases} y_{n_{k \bmod 4}} & \text{for } k = -n, \ldots, n-1, n \\ \begin{bmatrix} 0 \\ 0 \end{bmatrix} & \text{for } k \neq -n, \ldots, n-1, n \end{cases}$$

We see that, if λ is an almost eigenvalue of $\boldsymbol{Z}^*\boldsymbol{A}$, then $|\lambda|$ is also one, with a corresponding real almost eigenvector $\boldsymbol{y}_{1n}$ and/or $\boldsymbol{y}_{2n}$. Since $\|Re(\tilde{\boldsymbol{x}}_n)\|_2^2 + \|Im(\tilde{\boldsymbol{x}}_n)\|_2^2 = 1$, one of these two has a squared norm larger than or equal to $1/2$. By choosing this one, it is immediate that the resulting $\boldsymbol{y}_n$ satisfies

$$\|\boldsymbol{C}\boldsymbol{y}_n\|_2^2 \leq 2\|\boldsymbol{C}\boldsymbol{x}_n\|_2^2$$

Therefore, if we try to prove Conjecture 6.26 by assuming that the pair $(\boldsymbol{C}, \boldsymbol{A})$ is not uniformly observable, we only have to consider real almost eigenvalues and real almost eigenvectors.

7 Continuous-Time Entropy

As was shown in Chapter 4, the definition of entropy can be extended to discrete-time time-varying systems by means of the spectral factor of a related positive definite infinite-dimensional operator. In this chapter we consider the corresponding extension of the definition of entropy for continuous-time time-varying systems.

The extension of our notion of entropy to continuous-time systems is not straightforward. It is known that for continuous-time systems, there exist positive-definite operators which do not admit spectral factors [25]. To circumvent this difficulty we will restrict the class of systems that is considered in this chapter to the set of integral operators that have continuous kernels for which a spectral factorization exists of the same form.

The definition of entropy given in this chapter has been chosen to preserve most of the properties of the discrete-time entropy. This continuous-time entropy differs from the entropy for discrete-time time-varying systems of Chapter 4. In particular, the continuous-time entropy is not defined in terms of the memoryless part of the spectral factor. This difference is necessitated by the fact that, unlike discrete-time systems, a change in the input at a single time instant does not influence the output of the system.

7.1 Classes of systems considered

The notation we use is completely analogous to the notation that was used for discrete-time signals. The essential difference is that we consider signals belonging to $\mathcal{L}_2$, i.e. signals x_t satisfying

$$\|x\|_2 := \sqrt{\int_{-\infty}^{\infty} x_t^T x_t \, dt} < \infty$$

We will consider systems of a special form, namely those which can be represented as integral operators with continuous kernels. While not all LTV systems admit such a representation, this is a fairly large class of systems [72]. In order to define a finite entropy, we will restrict ourselves to strictly causal systems. For such a system G, we may write

$$(Gw)_t = \int_{-\infty}^{t} g_{t,\tau} w_\tau \, d\tau \tag{7.1}$$

As is well known, this class of systems includes all systems which can be represented by a state-space realization. Suppose that G is given by

$$\Sigma_G := \begin{cases} \dot{x}_t &= A_t x_t + B_t w_t \\ z_t &= C_t x_t \end{cases} \tag{7.2}$$

where all matrices are assumed to be uniformly bounded. Then we have that $g_{t,\tau} = C_t \Phi_{t,\tau} B_\tau$, where $\Phi_{t,\tau}$ is the transition matrix corresponding to A_t.

We say that A_t is uniformly exponentially stable (UES) if there exist c and $\beta > 0$ such that $\|\Phi_{t,\tau}\| \leq c e^{-\beta(t-\tau)}$ for all $t \geq \tau$. For a system G that admits a state-space realization, we say that G is stable if A_t is UES. By $\|G\|$ we denote the $\mathcal{L}_2$ induced norm of G. Obviously, for systems described by (7.2), $\|G\|$ is finite if A_t is UES.

A result that we will use repeatedly is the following:

Lemma 7.1 ([64]) *Let S be a square matrix, and let $\epsilon \in \mathbb{R}$. Then*

$$-\ln\det(I - \epsilon S) = \epsilon \operatorname{tr} S + \mathcal{O}(\epsilon^2)$$

In the next section we will define the entropy for systems described by (7.1).

7.2 Entropy of a continuous-time time-varying system

For a stable, continuous-time linear time-invariant system with transfer function $G(s)$ satisfying the norm constraint $\|G\|_\infty < \gamma$, the entropy at infinity is defined in [64] as

$$I_c(G,\gamma) := \lim_{s_0\to\infty} -\frac{\gamma^2}{2\pi}\int_{-\infty}^{\infty} \ln\left|\det\left(I-\gamma^{-2}G(-i\omega)^T G(i\omega)\right)\right| \frac{s_0^2}{s_0^2+\omega^2}\,d\omega \tag{7.3}$$

In [64] Mustafa and Glover showed that

$$I_c(G,\gamma) \geq \frac{1}{2\pi}\int_{-\infty}^{\infty} \mathrm{tr}[G(-i\omega)^T G(i\omega)]\,d\omega = \|G\|_2^2$$

The $\mathcal{H}_2$ norm in this inequality measures the influence of white noise on the output. If $z = Gw$ where w is the derivative of a Wiener process $(\mathcal{E}w_t w_s^T = I\delta_{t-s})$, then

$$\|G\|_2^2 = \lim_{T\to\infty} \mathcal{E}\frac{1}{2T}\int_{-T}^{T} |z_t|^2\,dt$$

An interpretation of entropy as an $\mathcal{H}_2$ cost for which large deviations from the mean of the output are penalized is outlined in [39], where it is shown that

$$I_c(G,\gamma) = \lim_{T\to\infty} -\frac{\gamma^2}{T}\ln\mathcal{E}\exp\Big(\frac{1}{2\gamma^2}\int_{-T}^{T}|z_t|^2\,dt\Big)$$

In Chapter 4, it was shown that for a discrete-time system, the LTI entropy can be extended to the time-varying case using a spectral factorization. For continuous-time systems it is known that spectral factorizations of a positive operator need not exist [25]. We will extend the entropy (7.3) to the class of time-varying systems described by an integral operator (7.1), for which a spectral factorization exists of the same form. It will be shown in Section 7.3.2 that this is the case for systems described by a state-space realization. Therefore, suppose that G is a strictly causal integral operator with continuous kernel, and $I-\gamma^{-2}G^*G$ is a positive operator which admits a spectral factorization

$$I-\gamma^{-2}G^*G = M^*M \tag{7.4}$$

where M is causal with a causal inverse. Moreover, assume that

$$(Mw)_t = \hat{m}_t w_t + \int_{-\infty}^{t} m_{t,\tau} w_\tau\,d\tau$$

where $m_{t,\tau}$ is continuous in both arguments. For completeness we mention that

$$\left((I-\gamma^{-2}G^*G)w\right)_t = w_t - \gamma^{-2}\int_t^\infty g_{\tau,t}^T \int_{-\infty}^{\tau} g_{\tau,\sigma} w_\sigma \, d\sigma d\tau \qquad (7.5)$$

The entropy will be defined in terms of the operator satisfying the spectral factorization (7.4).

Definition 7.2 *Suppose that G is strictly causal with $\|G\| < \gamma$, and that $I-\gamma^{-2}G^*G$ admits a spectral factorization (7.4). The entropy at time t is defined as*

$$E(G,\gamma,t) := \lim_{\alpha\downarrow 0} -\frac{\gamma^2}{2\alpha} \ln\det\left((\hat{m}_t + \alpha m_{t,t})^T(\hat{m}_t + \alpha m_{t,t})\right) \qquad (7.6)$$

We must first show that this definition makes sense. Note that since G is strictly causal, by taking $w_\tau = 0$ for $\tau \neq t$, we have that $\hat{m}_t^T\hat{m}_t = I$. Using Lemma 7.1 we can evaluate the entropy as

$$\begin{aligned} E(G,\gamma,t) &= \lim_{\alpha\downarrow 0} -\frac{\gamma^2}{2\alpha} \ln\det\left(I + \alpha\hat{m}_t^T m_{t,t} + \alpha m_{t,t}^T\hat{m}_t + \alpha^2 m_{t,t}^T m_{t,t}\right) \\ &= \lim_{\alpha\downarrow 0} -\frac{\gamma^2}{2} \operatorname{tr}\left[\hat{m}_t^T m_{t,t} + m_{t,t}^T\hat{m}_t + \alpha m_{t,t}^T m_{t,t}\right] + \mathcal{O}(\alpha) \\ &= -\frac{\gamma^2}{2}\operatorname{tr}\left[\hat{m}_t^T m_{t,t} + m_{t,t}^T\hat{m}_t\right] \end{aligned} \qquad (7.7)$$

which is well-defined. In [25] it is shown that if $\bar{M}$ is an other spectral factor, then $\bar{M} = UM$, where $U^*U = I$. Using this it is easy to see that the entropy does not depend on the particular spectral factor chosen.

Remark 7.3 Since $\hat{m}_t$ is unitary, we can choose $\hat{m}_t = I$ without loss of generality. This will be done throughout the sequel.

In the next section we will discuss some properties of the continuous-time time-varying entropy.

7.3 Properties

As we showed in Chapter 6 for discrete-time systems, it is essential for minimization purposes that the entropy be non-negative, and equal to zero if and only if the operator is zero. For the continuous-time entropy defined above, we now prove similar results.

Lemma 7.4 *Suppose that G is strictly causal with norm bound $\|G\| < \gamma$. Then we have that $E(G, \gamma, t) \geq 0$ for all t.*

Proof: For every $\epsilon > 0$ there exists $\nu > 0$ such that

$$\|m_{r,\sigma} - m_{t,t}\| < \epsilon \quad \text{for all } r \in [t-\nu, t] \text{ and } \sigma \in [t-\nu, r] \tag{7.8}$$

This can easily be seen using

$$\|m_{r,\sigma} - m_{t,t}\| \leq \|m_{r,\sigma} - m_{t,\sigma}\| + \|m_{t,\sigma} - m_{t,t}\|$$

and continuity arguments. Now, for any $\alpha \in [0, \nu)$, define a signal w such that $w_r = \bar{w}$ for $r \in [t-\alpha, t]$, and 0 otherwise, where $\bar{w}$ is a constant vector. We know that

$$\begin{aligned} \langle w, M^*Mw \rangle &= \langle w, w \rangle - \gamma^{-2} \langle w, G^*Gw \rangle \\ &\leq \langle w, w \rangle = \int_{t-\alpha}^{t} |\bar{w}|^2 \, dr = \alpha |\bar{w}|^2 \end{aligned} \tag{7.9}$$

On the other hand, we have

$$\begin{aligned} \langle w, M^*Mw \rangle &= \int_{-\infty}^{\infty} |(Mw)_r|^2 \, dr \\ &= \int_{-\infty}^{\infty} \left| w_r + \int_{-\infty}^{r} m_{r,\sigma} w_\sigma \, d\sigma \right|^2 dr \\ &\geq \int_{-\infty}^{\infty} |w_r|^2 \, dr + 2 \int_{-\infty}^{\infty} w_r^T \int_{-\infty}^{r} m_{r,\sigma} w_\sigma \, d\sigma \, dr \end{aligned} \tag{7.10}$$

where the inequality follows by leaving the quadratic term out. Using the definition of the input signal w_r, (7.10) can be written as

$$\begin{aligned} \langle w, M^*Mw \rangle &\geq \int_{t-\alpha}^{t} |\bar{w}|^2 \, dr + 2 \int_{t-\alpha}^{t} \bar{w}^T \int_{t-\alpha}^{r} m_{r,\sigma} \bar{w} \, d\sigma \, dr \\ &= \alpha |\bar{w}|^2 + 2 \int_{t-\alpha}^{t} \bar{w}^T \int_{t-\alpha}^{r} m_{t,t} \bar{w} \, d\sigma \, dr \\ &\quad + 2 \int_{t-\alpha}^{t} \bar{w}^T \int_{t-\alpha}^{r} (m_{r,\sigma} - m_{t,t}) \bar{w} \, d\sigma \, dr \end{aligned}$$

The second term can be evaluated as $\alpha^2 \bar{w}^T m_{t,t} \bar{w}$. Furthermore, using (7.8) it is easy to see that the third term is bounded below by $-\epsilon\alpha^2 |\bar{w}|^2$, yielding

$$\langle w, M^*Mw \rangle \geq \alpha |\bar{w}|^2 + \alpha^2 \bar{w}^T m_{t,t} \bar{w} - \epsilon \alpha^2 |\bar{w}|^2 \tag{7.11}$$

Combining inequalities (7.9) and (7.11) gives

$$\bar{w}^T m_{t,t}\bar{w} - \epsilon|\bar{w}|^2 \leq 0$$

Since ϵ is arbitrary, this shows that $\bar{w}^T m_{t,t}\bar{w} \leq 0$ for all $\bar{w}$. It follows that $m_{t,t} + m_{t,t}^T \leq 0$. ■

Lemma 7.5 *$E(G, \gamma, t) = 0$ for all t if and only if $G \equiv 0$.*

Proof: If $G \equiv 0$ it follows that $m_{t,\tau} = 0$ for all $t \geq \tau$ and the statement is obvious. Now suppose that $E(G, \gamma, t) = 0$ for all t. If $m_{t,\tau} = 0$ for all $t > \tau$, it follows by continuity that $m_{t,t} = 0$ for all t. Suppose that $m_{t,\tau} \neq 0$ for some $t > \tau$. For every $\epsilon > 0$, there exists $\nu > 0$ such that

$$\|m_{r,\sigma} - m_{t,t}\| < \epsilon \qquad \text{for all } r \in [t-\nu, t] \text{ and } \sigma \in [t-\nu, r] \tag{7.12}$$

$$\|m_{r,\sigma} - m_{\tau,\tau}\| < \epsilon \qquad \text{for all } r \in [\tau-\nu, \tau] \text{ and } \sigma \in [\tau-\nu, r] \tag{7.13}$$

$$\|m_{r,\sigma} - m_{t,\tau}\| < \epsilon \qquad \text{for all } r \in [t-\nu, t] \text{ and } \sigma \in [\tau-\nu, \tau] \tag{7.14}$$

all hold.

Since $m_{t,\tau} \neq 0$, there exist $\tilde{w}$ and $\bar{w}$ with $|\tilde{w}| = |\bar{w}| = 1$ such that $\tilde{w}^T m_{t,\tau}\bar{w} =: a > 0$. Since the entropy equals zero for all t, we know that $\bar{w}^T m_{\tau,\tau}\bar{w} = \tilde{w}^T m_{t,t}\tilde{w} = 0$. Let $\epsilon < \frac{a}{2}$. For any $\alpha \in [0, \nu)$ with $t - \alpha > \tau$ we define

$$w_r := \begin{cases} \bar{w} & \text{for } r \in [\tau-\alpha, \tau] \\ \tilde{w} & \text{for } r \in [t-\alpha, t] \\ 0 & \text{otherwise} \end{cases}$$

Using this input, we find as in (7.10)

$$\begin{aligned}
&\langle w, M^*Mw\rangle \\
&\geq \int_{-\infty}^{\infty} |w_r|^2\, dr + 2\int_{-\infty}^{\infty} w_r^T \int_{-\infty}^{r} m_{r,\sigma}w_\sigma\, d\sigma\, dr \\
&= \int_{\tau-\alpha}^{\tau} |\bar{w}|^2\, dr + \int_{t-\alpha}^{t} |\tilde{w}|^2\, dr + 2\int_{\tau-\alpha}^{\tau} \bar{w}^T \int_{\tau-\alpha}^{r} m_{r,\sigma}\bar{w}\, d\sigma\, dr \\
&\quad + 2\int_{t-\alpha}^{t} \tilde{w}^T \int_{\tau-\alpha}^{\tau} m_{r,\sigma}\bar{w}\, d\sigma\, dr + 2\int_{t-\alpha}^{t} \tilde{w}^T \int_{t-\alpha}^{r} m_{r,\sigma}\tilde{w}\, d\sigma\, dr
\end{aligned}$$

The first two terms both result in α. Since $m_{t,t} = 0$ for all t, this expression can be written as

$$\begin{aligned}
&\langle w, M^*Mw\rangle \\
&\geq 2\alpha + 2\int_{\tau-\alpha}^{\tau} \bar{w}^T \int_{\tau-\alpha}^{r} (m_{r,\sigma} - m_{\tau,\tau})\bar{w}\, d\sigma\, dr \\
&\quad + 2\int_{t-\alpha}^{t} \tilde{w}^T \int_{\tau-\alpha}^{\tau} m_{t,\tau}\bar{w}\, d\sigma\, dr + 2\int_{t-\alpha}^{t} \tilde{w}^T \int_{\tau-\alpha}^{\tau} (m_{r,\sigma} - m_{t,\tau})\bar{w}\, d\sigma\, dr \\
&\quad + 2\int_{t-\alpha}^{t} \tilde{w}^T \int_{t-\alpha}^{r} (m_{r,\sigma} - m_{t,t})\tilde{w}\, d\sigma\, dr
\end{aligned}$$

Using (7.12–7.14) the second and fifth term are both bounded below by $-\epsilon\alpha^2$, and the fourth one by $-2\epsilon\alpha^2$. Moreover, $\tilde{w}^T m_{t,\tau}\bar{w} = a$, resulting in

$$\begin{aligned}
\langle w, M^*Mw\rangle &\geq 2\alpha - \epsilon\alpha^2 + 2a\alpha^2 - 2\epsilon\alpha^2 - \epsilon\alpha^2 \\
&> 2\alpha
\end{aligned}$$

since $\epsilon < \frac{a}{2}$. On the other hand

$$\begin{aligned}
\langle w, M^*Mw\rangle &= \langle w, w\rangle - \gamma^{-2}\langle w, G^*Gw\rangle \\
&\leq \langle w, w\rangle = \int_{\tau-\alpha}^{\tau} |\bar{w}|^2\, dr + \int_{t-\alpha}^{t} |\bar{w}|^2\, dr = 2\alpha
\end{aligned}$$

which is a contradiction. ■

Lemma 7.6 *If $\gamma \geq \bar{\gamma} > \|G\|$, then $E(G,\gamma,t) \leq E(G,\bar{\gamma},t)$ for all t.*

Proof: Recall that M is the spectral factor of $I - \gamma^{-2}G^*G$. We can compute a spectral factor (say L) with respect to some other $\|G\| < \bar{\gamma} \leq \gamma$, giving

$$\gamma^2 I - \gamma^2 M^*M = G^*G = \bar{\gamma}^2 I - \bar{\gamma}^2 L^*L \tag{7.15}$$

For $\alpha > 0$, define the signal w according to

$$w_r = \begin{cases} 0 & \text{if } r < t - \alpha \\ \bar{w} & \text{if } r \in [t-\alpha, t] \end{cases}$$

where $\bar{w}$ is any unit length vector. For $r > t$ we take w_r satisfying

$$w_r = -\int_{-\infty}^{r} m_{r,\tau} w_\tau\, d\tau$$

such that $(Mw)_r = 0$ for $r > t$. Applying this w to (7.15) gives

$$\begin{aligned}
\gamma^2 \|w\|_2^2 - \gamma^2 \int_{t-\alpha}^{t} \left| \bar{w} + \int_{t-\alpha}^{r} m_{r,\tau} \bar{w}\, d\tau \right|^2 dr \\
= \bar{\gamma}^2 \|w\|_2^2 - \bar{\gamma}^2 \int_{t-\alpha}^{\infty} \left| w_r + \int_{t-\alpha}^{r} l_{r,\tau} w_\tau\, d\tau \right|^2 dr \\
\leq \bar{\gamma}^2 \|w\|_2^2 - \bar{\gamma}^2 \int_{t-\alpha}^{t} \left| w_r + \int_{t-\alpha}^{r} l_{r,\tau} w_\tau\, d\tau \right|^2 dr
\end{aligned}$$

Since

$$\|w\|_2^2 = \int_{t-\alpha}^{\infty} |w_r|^2\, dr = \alpha |\bar{w}|^2 + \int_{t}^{\infty} |w_r|^2\, dr$$

and $\bar{\gamma} \leq \gamma$, we have

$$\begin{aligned}
\gamma^2 \alpha |\bar{w}|^2 - \gamma^2 \int_{t-\alpha}^{t} \left| \bar{w} + \int_{t-\alpha}^{r} m_{r,\tau} \bar{w}\, d\tau \right|^2 dr \\
\leq \bar{\gamma}^2 \alpha |\bar{w}|^2 - \bar{\gamma}^2 \int_{t-\alpha}^{t} \left| w_r + \int_{t-\alpha}^{r} l_{r,\tau} w_\tau\, d\tau \right|^2 dr
\end{aligned} \tag{7.16}$$

Given an $\epsilon > 0$, there exists a $\nu > 0$ such that

$$\|m_{r,\tau} - m_{t,t}\| \leq \epsilon \quad \text{for all } r \in [t-\nu, t] \text{ and } \tau \in [t-\nu, r]$$

Using this we can evaluate the left-hand side of (7.16) as

$$\begin{aligned}
-2\gamma^2 \int_{t-\alpha}^{t} \bar{w}^T \int_{t-\alpha}^{r} m_{r,\tau} \bar{w}\, d\tau\, dr - \gamma^2 \int_{t-\alpha}^{t} \left| \int_{t-\alpha}^{r} m_{r,\tau} \bar{w}\, d\tau \right|^2 dr \\
= -\alpha^2 \gamma^2 \bar{w}^T m_{t,t} \bar{w} - 2\gamma^2 \int_{t-\alpha}^{t} \bar{w}^T \int_{t-\alpha}^{r} (m_{r,\tau} - m_{t,t}) \bar{w}\, d\tau\, dr \\
- \gamma^2 \int_{t-\alpha}^{t} \left| \int_{t-\alpha}^{r} m_{r,\tau} \bar{w}\, d\tau \right|^2 dr \\
= -\alpha^2 \gamma^2 \bar{w}^T m_{t,t} \bar{w} + \mathcal{O}(\epsilon \alpha^2) + \mathcal{O}(\alpha^3)
\end{aligned}$$

Similarly, we can evaluate the right-hand side of (7.16). Now, by dividing by α^2, and taking the limit as α goes to zero, gives

$$-\gamma^2 \bar{w}^T m_{t,t} \bar{w} \leq -\bar{\gamma}^2 \bar{w}^T l_{t,t} \bar{w}$$

since ϵ is arbitrary. Finally, since $\bar{w}$ is arbitrary, it follows that

$$\begin{aligned}
E(G, \gamma, t) = -\frac{\gamma^2}{2} \operatorname{tr}\left[m_{t,t} + m_{t,t}^T\right] \\
\leq -\frac{\bar{\gamma}^2}{2} \operatorname{tr}\left[l_{t,t} + l_{t,t}^T\right] = E(G, \bar{\gamma}, t)
\end{aligned}$$

using the expression (7.7) for the entropy. ■

7.3.1 Equivalence with the entropy integral

We will now evaluate the entropy from Definition 7.2 for a time-invariant system G, i.e. $g_{t,\tau} = g_{t-\tau}$. In this case, the LTI entropy (7.3) can be obtained from the entropy defined by (7.6) directly.

Lemma 7.7 *If G is time-invariant with $\|G\| < \gamma$, then for all t we have*

$$E(G, \gamma, t) = I_c(G, \gamma)$$

Proof: Since $\|G\| < \gamma$ there exists a spectral factorization

$$I - \gamma^{-2} G^* G = M^* M$$

It can be checked that

$$I - \gamma^{-2} G(s)^\sim G(s) = M(s)^\sim M(s)$$

where $M(s)$ and $G(s)$ are the Laplace transforms of M and G, and $G^\sim$ denotes the para-Hermitian conjugate $G(s)^\sim = G(-\bar{s})^H$. Since M is also time-invariant, we have that $m_{t,\tau} = m_{t-\tau}$, hence we can compute using $G(i\omega)^\sim = G(-i\omega)^T$

$$\begin{aligned} I_c(G,\gamma) &= \lim_{s_0 \to \infty} -\frac{\gamma^2}{2\pi} \int_{-\infty}^{\infty} \ln \left|\det\left(I - \gamma^{-2} G(-i\omega)^T G(i\omega)\right)\right| \frac{s_0^2}{s_0^2 + \omega^2}\, d\omega \\ &= \lim_{s_0 \to \infty} -\frac{\gamma^2 s_0}{\pi} \int_{-\infty}^{\infty} \ln \left|\det\left(M(i\omega)\right)\right| \frac{s_0}{s_0^2 + \omega^2}\, d\omega \\ &= \lim_{s_0 \to \infty} -\gamma^2 s_0 \ln \left|\det\left(I + \frac{m_0}{s_0} + \mathcal{O}\left(\frac{1}{s_0^2}\right)\right)\right| \end{aligned}$$

where the third equality follows from the continuous-time version of Poisson's integral formula. Substituting $s_0 = \frac{1}{\alpha}$ results in

$$\begin{aligned} I_c(G,\gamma) &= \lim_{\alpha \downarrow 0} -\frac{\gamma^2}{\alpha} \ln \left|\det\left(I + \alpha m_0 + \mathcal{O}(\alpha^2)\right)\right| \\ &= \lim_{\alpha \downarrow 0} -\frac{\gamma^2}{\alpha} \ln \left|\det\left(I + \alpha m_0\right)\right| \\ &= \lim_{\alpha \downarrow 0} -\frac{\gamma^2}{2\alpha} \ln \det\left((I + \alpha m_{t,t})^T (I + \alpha m_{t,t})\right) \end{aligned}$$

which is the entropy (7.6). ∎

7.3.2 Entropy in terms of a state-space realization

For systems which admit a state-space realization the entropy can be expressed in terms of the state-space matrices. Suppose that G is given by (7.2) with A_t UES. From [67] we know that $\|G\| < \gamma$ if and only if there exists a positive semi-definite bounded solution X_t satisfying

$$-\dot{X}_t = A_t^T X_t + X_t A_t + C_t^T C_t + \gamma^{-2} X_t B_t B_t^T X_t \tag{7.17}$$

such that $A_t + \gamma^{-2} B_t B_t^T X_t$ is UES. Using the solution to this Riccati equation we can find a spectral factor of $I - \gamma^{-2} G^* G$.

Lemma 7.8 *For a stable system satisfying $\|G\| < \gamma$, where G is given by a state-space realization (7.2), a spectral factorization $I - \gamma^{-2} G^* G = M^* M$ is given by*

$$\Sigma_M := \begin{cases} \dot{x}_t = A_t x_t + B_t w_t \\ z_t = -\gamma^{-2} B_t^T X_t x_t + w_t \end{cases}$$

Proof: We need to show that both M and M^{-1} are causal, bounded operators on $\mathcal{L}_2$; and that $I - \gamma^{-2} G^* G = M^* M$ is satisfied. We first prove the latter.

Using the kernel $g_{t,\tau} = C_t \Phi_{t,\tau} B_\tau$, we find that, for any signal $w \in \mathcal{L}_2$

$$\begin{aligned} &\left((I - \gamma^{-2} G^* G) w\right)_t \\ &\quad = w_t - \gamma^{-2} \int_t^\infty B_t^T \Phi_{\tau,t}^T C_\tau^T \int_{-\infty}^\tau C_\tau \Phi_{\tau,r} B_r w_r \, dr \, d\tau \end{aligned}$$

Using the Riccati equation (7.17), we get

$$\begin{aligned} &\left((I - \gamma^{-2} G^* G) w\right)_t \\ &\quad = w_t + \gamma^{-4} \int_t^\infty B_t^T \Phi_{\tau,t}^T X_\tau B_\tau B_\tau^T X_\tau \int_{-\infty}^\tau \Phi_{\tau,r} B_r w_r \, dr \, d\tau \\ &\qquad + \gamma^{-2} \int_t^\infty B_t^T \Phi_{\tau,t}^T (\dot{X}_\tau + A_t^T X_\tau + X_\tau A_\tau) \int_{-\infty}^\tau \Phi_{\tau,r} B_r w_r \, dr \, d\tau \\ &\quad = w_t + \gamma^{-4} \int_t^\infty B_t^T \Phi_{\tau,t}^T X_\tau B_\tau B_\tau^T X_\tau \int_{-\infty}^\tau \Phi_{\tau,r} B_r w_r \, dr \, d\tau \\ &\qquad + \gamma^{-2} \int_t^\infty B_t^T \int_{-\infty}^\tau \frac{\partial}{\partial \tau} [\Phi_{\tau,t}^T X_\tau \Phi_{\tau,r}] B_r w_r \, dr \, d\tau \end{aligned}$$

Changing the order of integration yields

$$
\begin{aligned}
&\left((I-\gamma^{-2}G^*G)w\right)_t \\
&= w_t + \gamma^{-4}\int_t^{\infty} B_t^T\Phi_{\tau,t}^T X_\tau B_\tau B_\tau^T X_\tau \int_{-\infty}^{\tau} \Phi_{\tau,r}B_r w_r\, dr\, d\tau \\
&\quad + \gamma^{-2}\int_{-\infty}^{t}\int_t^{\infty} B_t^T \frac{\partial}{\partial \tau}\left[\Phi_{\tau,t}^T X_\tau \Phi_{\tau,r}\right] B_r w_r\, d\tau\, dr \\
&\quad + \gamma^{-2}\int_t^{\infty}\int_r^{\infty} B_t^T \frac{\partial}{\partial \tau}\left[\Phi_{\tau,t}^T X_\tau \Phi_{\tau,r}\right] B_r w_r\, d\tau\, dr \\
&= w_t + \gamma^{-4}\int_t^{\infty} B_t^T\Phi_{\tau,t}^T X_\tau B_\tau B_\tau^T X_\tau \int_{-\infty}^{\tau} \Phi_{\tau,r}B_r w_r\, dr\, d\tau \\
&\quad - \gamma^{-2}\int_{-\infty}^{t} B_t^T X_t \Phi_{t,r} B_r w_r\, dr - \gamma^{-2}\int_t^{\infty} B_t^T \Phi_{r,t}^T X_r B_r w_r\, dr \\
&= (M^*Mw)_t
\end{aligned}
$$

where we used that $\lim_{\tau\to\infty} \Phi_{\tau,t} = 0$

That M is a causal bounded operator is immediate from the state-space realization of M, and the fact that for a system with bounded state-space matrices, UES implies $\mathcal{L}_2$ stability. This fact also ensures that M^{-1} is a causal bounded operator since M^{-1} has a state-space realization:

$$
\Sigma_{M^{-1}} := \left\{ \begin{array}{rcl} \dot{\xi}_t &=& A_t + \gamma^{-2}B_tB_t^TX_t\xi_t + B_t\hat{w}_t \\ \hat{z}_t &=& \gamma^{-2}B_t^TX_t\xi_t + \hat{w}_t \end{array} \right.
$$

where all of the matrices are bounded, and $A_t + \gamma^{-2}B_tB_t^TX_t$ is UES. ■

Since $m_{t,\tau} = -\gamma^{-2}B_t^TX_t\Phi_{t,\tau}B_\tau$, an expression for the entropy follows immediately.

Corollary 7.9 *For a stable system satisfying $\|G\| < \gamma$, where G is given by a state-space realization (7.2), the entropy is given by*

$$
E(G,\gamma,t) = \operatorname{tr}\left[B_t^TX_tB_t\right]
$$

where X is the stabilizing solution to (7.17). □

This state-space expression is analogous to that given for time-invariant systems in [64].

Remark 7.10 Since the solution to (7.17) is a decreasing function of γ (uniformly in t), it follows from this expression for the entropy that the entropy is a decreasing function of γ, in accordance with Lemma 7.6.

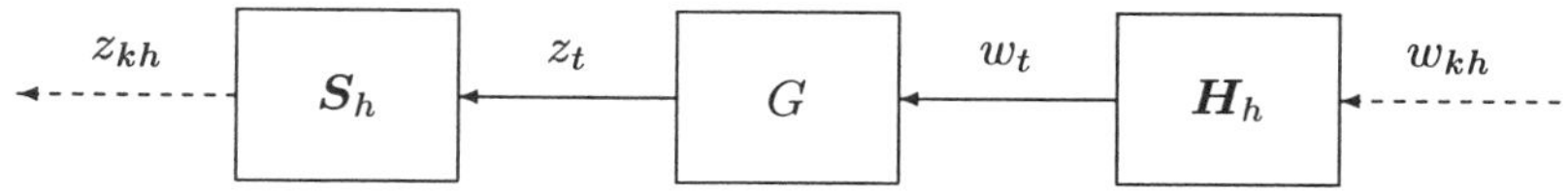

Figure 7.1: Hold and sample system

Remark 7.11 This expression for the entropy is equal to the so-called auxiliary cost [15]. This auxiliary cost has been used to minimize an upper bound for the $\mathcal{H}_2$ cost among all controllers satisfying an $\mathcal{H}_\infty$ constraint.

7.3.3 Relationship with discrete-time entropy

We wish to relate the definition of entropy for continuous-time systems to the entropy for discrete-time systems. We will do so using by considering appropriate sample and hold operators.

We begin by considering a discrete-time sequence as the input to a hold operator. The resulting continuous-time signal is used as the input to the continuous-time system. The output of this continuous-time system is then sampled using the ideal sampling operator. The cascade of these three systems is illustrated in Figure 7.1 where h is the clock time in the sample and hold circuits.

Take any sequence $\{v_k\} \in \ell_2^m$. We define the continuous-time signal

$$w_t = v_k, \quad t \in [kh, (k+1)h)$$

For this input, the sampled output equals

$$z_{kh} = \int_{-\infty}^{kh} g_{kh,\tau} w_\tau d\tau = \sum_{i=-\infty}^{k-1} G_{k,i} v_i$$

where

$$G_{k,i} := \int_{ih}^{(i+1)h} g_{kh,\tau} d\tau$$

The corresponding discrete-time operator will be denoted by $\boldsymbol{G}$, so that $\boldsymbol{G} := \boldsymbol{S}_h G \boldsymbol{H}_h$. Note that, for fixed h, computing the corresponding operator $\boldsymbol{M} := \boldsymbol{S}_h G \boldsymbol{H}_h$ for the continuous-time spectral factor M does *not* result in a spectral factorization of the discrete-time operator $\boldsymbol{I} - \gamma^{-2}\boldsymbol{G}^*\boldsymbol{G}$.

Instead, we compute a factorization of this positive-definite operator and denote it by $\boldsymbol{L}$; i.e.

$$\boldsymbol{I} - \gamma^{-2}\boldsymbol{G}^*\boldsymbol{G} = \boldsymbol{L}^*\boldsymbol{L} \tag{7.18}$$

In order to discuss the relationship between the continuous-time entropy and the entropy for discrete-time systems as defined in Chapter 4, we will use a result on the norm of hybrid continuous-time time-varying systems.

Lemma 7.12 **([50])** *Define* $s_{t,\tau}(h)$ *by*

$$s_{t,\tau}(h) := \sup_{0<\delta<h} \|g_{kh+\delta,\tau} - g_{kh,\tau}\|, \quad kh < t \leq (k+1)h, \quad \text{and } \tau < t$$

Suppose that

(i) $\lim_{h\downarrow 0} \sup_{0<\epsilon<\delta<h} \|g_{kh+\delta,kh+\epsilon}\| < \infty$, *for all* k

(ii) $\lim_{h\downarrow 0} \int_{-\infty}^{\infty} |s_{t,\tau}(h)|\, d\tau = 0$, *for all* t

(iii) $\lim_{h\downarrow 0} \int_{-\infty}^{\infty} |s_{t,\tau}(h)|\, dt = 0$, *for all* τ

Then $\|(I - \boldsymbol{H}_h\boldsymbol{S}_h)G\|$ *converges to zero as* h *goes to zero.* □

Remark 7.13 The assumptions are very technical. For a state-space realization (7.2), where A_t is UES and the state-space matrices are continuous and uniformly bounded functions of t, these assumptions are satisfied if the derivative of C_t is uniformly bounded.

Using the *floor* function, defined for $a \in \mathbb{R}$ by

$$\lfloor a \rfloor := b \quad \text{where } b \in \mathbb{Z} \text{ such that } b \leq a < b+1$$

we are ready to show the following:

Theorem 7.14 *Suppose that* G *is strictly causal with* $\|G\| < \gamma$. *If* G *satisfies the assumptions in Lemma 7.12, then*

$$E(G, \gamma, t) = \lim_{h\downarrow 0} \frac{1}{h} \boldsymbol{E}(\boldsymbol{S}_h G \boldsymbol{H}_h, \gamma)_{\lfloor t/h \rfloor}$$

Proof: From Lemma 2.3 we know that the memoryless part $\boldsymbol{L}_m$ is invertible. Hence we can define the input v_i by

$$v_i = \begin{cases} 0 & \text{if } i < k \\ \bar{v} & \text{if } i = k \\ -L_{i,i}^{-1}\sum_{j=k}^{i-1} L_{i,j}v_j & \text{if } i > k \end{cases} \tag{7.19}$$

where $\bar{v}$ is any unit length vector. Notice that $(Lv)_i = 0$ for $i > k$. It follows from (7.18) that

$$\sum_{j=k}^{\infty} |v_j|^2 - \gamma^{-2} \sum_{j=k+1}^{\infty} \left|(\boldsymbol{G}\boldsymbol{w})_j\right|^2 = \left|L_{k,k}w_k\right|^2 \tag{7.20}$$

On the other hand, by construction of $\boldsymbol{G}$, we know that [1]

$$\begin{aligned} &\|\boldsymbol{w}\|_{\ell_2}^2 - \gamma^{-2}\|\boldsymbol{G}\boldsymbol{w}\|_{\ell_2}^2 \\ &\quad = \frac{1}{h}\|\boldsymbol{H}_h\boldsymbol{w}\|_{\mathcal{L}_2}^2 - \gamma^{-2}\frac{1}{h}\|\boldsymbol{H}_h\boldsymbol{S}_h G\boldsymbol{H}_h\boldsymbol{w}\|_{\mathcal{L}_2}^2 \\ &\quad = \frac{1}{h}\|\boldsymbol{H}_h\boldsymbol{w}\|_{\mathcal{L}_2}^2 - \gamma^{-2}\frac{1}{h}\|G\boldsymbol{H}_h\boldsymbol{w}\|_{\mathcal{L}_2}^2 \\ &\qquad + \gamma^{-2}\frac{1}{h}\|G\boldsymbol{H}_h\boldsymbol{w}\|_{\mathcal{L}_2}^2 - \gamma^{-2}\frac{1}{h}\|\boldsymbol{H}_h\boldsymbol{S}_h G\boldsymbol{H}_h\boldsymbol{w}\|_{\mathcal{L}_2}^2 \end{aligned} \tag{7.21}$$

Consider the last two terms of this expression. We have

$$\begin{aligned} &\left|\|G\boldsymbol{H}_h\boldsymbol{w}\|_{\mathcal{L}_2}^2 - \|\boldsymbol{H}_h\boldsymbol{S}_h G\boldsymbol{H}_h\boldsymbol{w}\|_{\mathcal{L}_2}^2\right| \\ &\quad = \left|\langle G\boldsymbol{H}_h\boldsymbol{w} - \boldsymbol{H}_h\boldsymbol{S}_h G\boldsymbol{H}_h\boldsymbol{w}, G\boldsymbol{H}_h\boldsymbol{w} + \boldsymbol{H}_h\boldsymbol{S}_h G\boldsymbol{H}_h\boldsymbol{w}\rangle_{\mathcal{L}_2}\right| \\ &\quad \le \|G - \boldsymbol{H}_h\boldsymbol{S}_h G\|\,\|\boldsymbol{H}_h\boldsymbol{w}\|_{\mathcal{L}_2}\,\|G\boldsymbol{H}_h\boldsymbol{w} + \boldsymbol{H}_h\boldsymbol{S}_h G\boldsymbol{H}_h\boldsymbol{w}\|_{\mathcal{L}_2} \end{aligned}$$

From Lemma 7.12 it follows, under the stated assumptions, that

$$\lim_{h\downarrow 0}\frac{1}{h}\|G - \boldsymbol{H}_h\boldsymbol{S}_h G\|\,\|\boldsymbol{H}_h\boldsymbol{w}\|_{\mathcal{L}_2} = 0$$

Hence, from (7.21)

$$\lim_{h\downarrow 0}\|\boldsymbol{w}\|_{\ell_2}^2 - \gamma^{-2}\|\boldsymbol{G}\boldsymbol{w}\|_{\ell_2}^2 = \lim_{h\downarrow 0}\frac{1}{h}\|\boldsymbol{H}_h\boldsymbol{w}\|_{\mathcal{L}_2}^2 - \gamma^{-2}\frac{1}{h}\|G\boldsymbol{H}_h\boldsymbol{w}\|_{\mathcal{L}_2}^2 \tag{7.22}$$

[1] The subscripts ℓ_2 and $\mathcal{L}_2$ emphasize that we are dealing with discrete-time respectively continuous-time signals.

Substituting this in (7.20) gives

$$\begin{aligned}\lim_{h\downarrow 0}\left|L_{k,k}\bar{v}\right|^2 &= \lim_{h\downarrow 0}\frac{1}{h}\left\|\boldsymbol{H}_h\boldsymbol{w}\right\|_{\mathcal{L}_2}^2 - \gamma^{-2}\frac{1}{h}\left\|G\boldsymbol{H}_h\boldsymbol{w}\right\|_{\mathcal{L}_2}^2\\ &= \lim_{h\downarrow 0}\frac{1}{h}\left\|M\boldsymbol{H}_h\boldsymbol{w}\right\|_{\mathcal{L}_2}^2\end{aligned}$$

The input $\boldsymbol{v}\in\ell_2^m$ in (7.19) is such that, given v_i being zero for $i<k$ and being equal to $\bar{v}$ for $i=k$, the cost function $\|\boldsymbol{v}\|_2^2-\gamma^{-2}\|\boldsymbol{G}\boldsymbol{v}\|_2^2$ is minimized. This is because $(\boldsymbol{L}\boldsymbol{v})_i=0$ for $i>k$. As h goes to zero, this cost function is related to the corresponding continuous-time one by (7.22).

For fixed h, and given w_r for $r\le t$, the minimum value of this cost function is obtained by taking $w_r=-\int_{-\infty}^{r} m_{r,\tau}w_\tau\,d\tau$ for $r>t$, such that $(Mw)_r=0$ for $r>t$. This results in

$$\begin{aligned}\lim_{h\downarrow 0}\left|L_{k,k}\bar{v}\right|^2 &= \lim_{h\downarrow 0}\frac{1}{h}\left\|M\boldsymbol{H}_h\boldsymbol{v}\right\|_{\mathcal{L}_2}^2\\ &= \lim_{h\downarrow 0}\frac{1}{h}\int_{kh}^{(k+1)h}\left|\bar{v}+\int_{kh}^{t} m_{t,\tau}\,d\tau\,\bar{v}\right|^2 dt\\ &= |\bar{v}|^2+\lim_{h\downarrow 0}\frac{1}{h}\bar{v}^T m_{kh,kh}\bar{v}\end{aligned}$$

where we used continuity of the kernel $m_{t,\tau}$. Since $\bar{v}$ is arbitrary, we can relate the memoryless part of L to the memoryless part of M by

$$\lim_{h\downarrow 0}L_{k,k}^T L_{k,k} = I+\lim_{h\downarrow 0}\frac{1}{h}\left(\frac{1}{2}m_{kh,kh}+\frac{1}{2}m_{kh,kh}^T\right)$$

By Definition 4.1, the (scaled) entropy corresponding to $\boldsymbol{L}$ at time k equals $-\frac{\gamma^2}{h}\ln\det(L_{k,k}^T L_{k,k})$. Hence, as the sampling period tends to zero, we get

$$\begin{aligned}E(G,\gamma,t) &= \lim_{h\downarrow 0}-\frac{\gamma^2}{h}\ln\det(L_{k,k}^T L_{k,k})\\ &= \lim_{h\downarrow 0}-\frac{\gamma^2}{h}\ln\det\left(I+\frac{1}{2h}\left(m_{kh,kh}+m_{kh,kh}^T\right)\right)\\ &= \lim_{h\downarrow 0}-\frac{\gamma^2}{2h}\ln\det\left(I+\frac{1}{h}\left(m_{kh,kh}+m_{kh,kh}^T\right)\right)\end{aligned}$$

The result follows by taking $k=\lfloor t/h\rfloor$. ■

In the next section we will show the relationships between minimum entropy control and $\mathcal{H}_\infty$, $\mathcal{H}_2$, and risk-sensitive control.

7.4 Connections with related optimal control problems

The connection between minimum entropy controllers and mixed $\mathcal{H}_2/\mathcal{H}_\infty$ controllers has been established in the LTI case [9]. Using the time-varying entropy introduced in Section 7.2, we now consider the corresponding relationship between the entropy with respect to $\mathcal{H}_2$ and $\mathcal{H}_\infty$ performance in the continuous-time case. The ideas are similar to those in Chapter 5 for the discrete-time case, although they are more straightforward in that setting since a different input at a single time instance changes the behavior of the system. For continuous-time systems this is not the case. Nevertheless, we can find a suitable measure for indicating the influence of the input at a certain time on the output.

7.4.1 Relationship with $\mathcal{H}_\infty$ control

$\mathcal{H}_\infty$ sub-optimal control is related to the cost function $\|w\|_2^2 - \gamma^{-2}\|z\|_2^2$. This cost function is non-negative for all inputs w if $\|G\| < \gamma$, and equals zero if and only if $w \equiv 0$. This is the so-called worst-case disturbance. We will investigate the guaranteed increase of this cost function for inputs $w \neq 0$.

Obviously, if w differs from zero on a set of measure zero, the cost will be unaffected. For this reason we consider signals w which are non-zero on a finite interval $[t-\alpha, t]$. During this time period, the signal will be constant equal to a unit vector $\bar{w}$. Finally, we will scale the cost function by dividing by α. Before proceeding, we introduce the continuous-time projection operator P_t defined by

$$(P_t w)_r := \begin{cases} w_r & \text{if } r \leq t \\ 0 & \text{if } r > t \end{cases}$$

and its orthogonal complement

$$(P_t^\perp w)_r := \begin{cases} 0 & \text{if } r \leq t \\ w_r & \text{if } r > t \end{cases}$$

We define the cost function $J(\gamma, t, \alpha)$ as

$$\frac{1}{\alpha} \inf_{P_t^\perp w} \left\{ \|w\|_2^2 - \gamma^{-2}\|z\|_2^2 \;\middle|\; P_{t-\alpha} w = 0,\; w_r = \bar{w} \text{ for } r \in [t-\alpha, t],\; |\bar{w}| = 1 \right\}$$

Obviously this function depends on the choice of $\bar{w}$. Nevertheless, since we are interested in the influence of an arbitrary input on $[t-\alpha, t]$, we would like to find a good bound for this function independent of $\bar{w}$. This we will do next; by scaling this function we can relate it to the entropy.

Lemma 7.15 *Suppose that G is strictly causal, and satisfies $\|G\| < \gamma$. Then*

$$0 \leq \lim_{\alpha \downarrow 0} -\frac{\gamma^2}{\alpha} \ln J(\gamma, t, \alpha) \leq E(G, \gamma, t) \tag{7.23}$$

Proof: Since $\|w\|_2^2 - \gamma^{-2}\|z\|_2^2 \leq \|w\|_2^2$, for which the infimum is attained by taking $P_t^{\perp} w = 0$, we get

$$J(\gamma, t, \alpha) \leq \frac{1}{\alpha} \int_{t-\alpha}^{t} |\bar{w}|^2 \, dt = 1$$

It follows that the function in (7.23) is non-negative. Now we want to show the upper-bound. Let $\epsilon > 0$. There exists a $\nu > 0$ such that

$$\|m_{r,\tau} - m_{t,t}\| \leq \epsilon \quad \text{for all } r \in [t-\nu, t] \text{ and } \tau \in [t-\nu, r] \tag{7.24}$$

Since $I - \gamma^{-2} G^* G = M^* M$, we get, given that $P_{r-\alpha} w = 0$ and $w_r = \bar{w}$ with $|\bar{w}| = 1$ for $r \in [t-\alpha, t]$

$$\begin{aligned}
J(\gamma, t, \alpha) &= \frac{1}{\alpha} \inf_{P_t^{\perp} w} \left\{ \int_{-\infty}^{\infty} \left| w_r + \int_{-\infty}^{r} m_{r,\tau} w_\tau \, d\tau \right|^2 dr \right\} \\
&= \frac{1}{\alpha} \int_{t-\alpha}^{t} \left| \bar{w} + \int_{t-\alpha}^{r} m_{r,\tau} \bar{w} \, d\tau \right|^2 dr \\
&= \frac{1}{\alpha} \int_{t-\alpha}^{t} \left| \bar{w} + \int_{t-\alpha}^{r} m_{t,t} \bar{w} \, d\tau + \int_{t-\alpha}^{r} (m_{r,\tau} - m_{t,t}) \bar{w} \, d\tau \right|^2 dr
\end{aligned}$$

where the infimum is obtained by choosing $w_r = -\int_{-\infty}^{r} m_{r,\tau} w_\tau \, d\tau$ for $r > t$. By splitting up the terms this can be evaluated as

$$\begin{aligned}
J(\gamma, t, \alpha) &= \frac{1}{\alpha} \int_{t-\alpha}^{t} \left| \bar{w} + (r - t + \alpha) m_{t,t} \bar{w} \, d\tau + \int_{t-\alpha}^{r} (m_{r,\tau} - m_{t,t}) \bar{w} \, d\tau \right|^2 dr \\
&= \frac{1}{\alpha} \int_{t-\alpha}^{t} \left| \left(I + (r - t + \alpha) m_{t,t} \right) \bar{w} \right|^2 dr \\
&\quad + \frac{1}{\alpha} \int_{t-\alpha}^{t} \left| \int_{t-\alpha}^{r} (m_{r,\tau} - m_{t,t}) \bar{w} \, d\tau \right|^2 dr \\
&\quad + \frac{2}{\alpha} \int_{t-\alpha}^{t} \bar{w}^T \left(I + (r - t + \alpha) m_{t,t} \right)^T \int_{t-\alpha}^{r} (m_{r,\tau} - m_{t,t}) \bar{w} \, d\tau \, dr
\end{aligned}$$

The first term can be computed exactly. Using (7.24) the second term can be shown to be of order $\alpha^2\epsilon^2$, yielding

$$\begin{aligned} J(\gamma,t,\alpha) &= \bar{w}^T(I+\alpha m_{t,t})\bar{w} + \frac{\alpha^2}{3}\left|m_{t,t}\bar{w}\right|^2 + \mathcal{O}(\alpha^2\epsilon^2) \\ &\quad + \frac{2}{\alpha}\int_{t-\alpha}^{t}\int_{t-\alpha}^{r} \bar{w}^T(m_{r,\tau}-m_{t,t})\bar{w}\,d\tau dr \\ &\quad + \frac{2}{\alpha}\int_{t-\alpha}^{t}\int_{t-\alpha}^{r} (r-t+\alpha)\bar{w}^T m_{t,t}^T(m_{r,\tau}-m_{t,t})\bar{w}\,d\tau dr \\ &= \bar{w}^T(I+\alpha m_{t,t})\bar{w} + \mathcal{O}(\alpha^2) + \mathcal{O}(\alpha^2\epsilon^2) + \mathcal{O}(\alpha\epsilon) + \mathcal{O}(\alpha^2\epsilon) \end{aligned}$$

where the last step again follows from (7.24). Since ϵ is arbitrary, it is easy to show that the limit in (7.23) exists, and equals

$$\lim_{\alpha\downarrow 0} -\frac{\gamma^2}{\alpha}\ln J(\gamma,t,\alpha) = \lim_{\alpha\downarrow 0} -\frac{\gamma^2}{\alpha}\ln\left(\bar{w}^T(I+\alpha m_{t,t})\bar{w}\right)$$

Since $m_{t,t}+m_{t,t}^T \le 0$ (Lemma 7.4), it follows that $0 < I+\frac{\alpha}{2}m_{t,t}+\frac{\alpha}{2}m_{t,t}^T \le I$ for α small enough. Hence it is standard mathematics to get

$$\begin{aligned} \lim_{\alpha\downarrow 0} -\frac{\gamma^2}{\alpha}\ln J(\gamma,t,\alpha) &= \lim_{\alpha\downarrow 0} -\frac{\gamma^2}{\alpha}\ln\left(\bar{w}^T(I+\frac{\alpha}{2}m_{t,t}+\frac{\alpha}{2}m_{t,t}^T)\bar{w}\right) \\ &\le \lim_{\alpha\downarrow 0} -\frac{\gamma^2}{\alpha}\ln\det\left(I+\frac{\alpha}{2}m_{t,t}+\frac{\alpha}{2}m_{t,t}^T\right) \end{aligned}$$

Now, using Lemma 7.1 we can evaluate this as

$$\begin{aligned} \lim_{\alpha\downarrow 0} -\frac{\gamma^2}{\alpha}\ln J(\gamma,t,\alpha) &\le \lim_{\alpha\downarrow 0} -\gamma^2\left(\operatorname{tr}\left[\frac{1}{2}m_{t,t}+\frac{1}{2}m_{t,t}^T\right]+\mathcal{O}(\alpha)\right) \\ &= -\frac{\gamma^2}{2}\operatorname{tr}\left[m_{t,t}+m_{t,t}^T\right] \\ &= E(G,\gamma,t) \end{aligned}$$

as required. ∎

Remark 7.16 Using similar arguments we can express the cost function in terms of the entropy by

$$1 \ge \lim_{\alpha\downarrow 0} J(\gamma,t,\alpha)^{1/\alpha} \ge \exp\left(-\gamma^{-2}E(G,\gamma,t)\right)$$

Hence all $\mathcal{H}_\infty$ controllers which ensure that the norm bound $\|G\| < \gamma$ is satisfied, assure that $\|w\|_2^2 - \gamma^{-2}\|z\|_2^2$ is larger than or equal to zero for all $0 \neq w \in \mathcal{L}_2$. This value equals 0 if the input is the worst-case disturbance $w \equiv 0$. The cost function $J(\gamma, t, \alpha)$ is an indication for the influence of w at time t. Within the set of γ sub-optimal controllers, the minimum entropy controller assures that the cost function increases with a guaranteed amount whenever the input w is not the worst-case input. Moreover, the increase achieved in the cost function is higher than the increase that other controllers guarantee.

7.4.2 Relationship with $\mathcal{H}_2$ control

Instead of taking the standard semi-norm

$$\limsup_{T\to\infty} \frac{1}{2T} \int_{-T}^{T} \int_{-T}^{\tau} \operatorname{tr}\left[g_{\tau,r}^T g_{\tau,r}\right] dr\, d\tau$$

we take an operator form for expressing the quadratic cost. Similar to the discrete-time case (see (5.5)), we define

$$Q(G,t) := \int_t^\infty \operatorname{tr}\left[g_{\tau,t}^T g_{\tau,t}\right] d\tau$$

The interpretation of this quadratic cost is pretty straightforward, as indicated by the following lemma. Instead of working with stochastic differential equations, we will regard w as white noise. Hence suppose that w is white noise, with $\mathcal{E} w_t w_s^T = I\delta_{t-s}$. Then we get the following:

Lemma 7.17 *Let w be white noise on $[t-\alpha, t]$, and zero elsewhere. Then*

$$Q(G,t) = \lim_{\alpha\downarrow 0} \frac{1}{\alpha} \mathcal{E}\, \|z\|_2^2$$

Proof: Since $z = Gw$, we get for any $\alpha > 0$

$$\begin{aligned}
\mathcal{E}\, \|z\|_2^2 &= \mathcal{E}\langle w, G^* G w\rangle \\
&= \mathcal{E} \int_{-\infty}^{\infty} w_r^T \int_r^\infty g_{\tau,r}^T \int_{-\infty}^{\tau} g_{\tau,\sigma} w_\sigma \, d\sigma\, d\tau\, dr
\end{aligned}$$

Since the input is zero everywhere outside the interval $[t-\alpha, t]$, this can be simplified as

$$\begin{aligned}
\mathcal{E}\, \|z\|_2^2 &= \mathcal{E} \int_{t-\alpha}^{t} w_r^T \int_r^\infty g_{\tau,r}^T \int_{-\infty}^{\tau} g_{\tau,\sigma} w_\sigma \, d\sigma\, d\tau\, dr \\
&= \mathcal{E} \int_{t-\alpha}^{t} w_r^T \int_r^t g_{\tau,r}^T \int_{t-\alpha}^{\tau} g_{\tau,\sigma} w_\sigma \, d\sigma\, d\tau\, dr \\
&\quad + \mathcal{E} \int_{t-\alpha}^{t} w_r^T \int_t^\infty g_{\tau,r}^T \int_{t-\alpha}^{t} g_{\tau,\sigma} w_\sigma \, d\sigma\, d\tau\, dr
\end{aligned}$$

It is easily seen that the first term is of order α^3. By changing the order of integration the second term can be evaluated as

$$\begin{aligned}
&\mathcal{E}\int_t^\infty\int_{t-\alpha}^t\int_{t-\alpha}^t w_r^T g_{\tau,r}^T g_{\tau,\sigma} w_\sigma \, d\sigma\, dr\, d\tau \\
&\qquad = \mathcal{E}\int_t^\infty\int_{t-\alpha}^t\int_{t-\alpha}^t \operatorname{tr}\left[g_{\tau,\sigma} w_\sigma w_r^T g_{\tau,r}^T\right] d\sigma\, dr\, d\tau \\
&\qquad = \int_t^\infty\int_{t-\alpha}^t\int_{t-\alpha}^t \operatorname{tr}\left[g_{\tau,\sigma} g_{\tau,r}^T\right]\delta_{r-\sigma}\, d\sigma\, dr\, d\tau
\end{aligned}$$

since w is white noise on $[t-\alpha, t]$. Now, by definition of the delta function, this yields

$$\begin{aligned}
&\mathcal{E}\int_t^\infty\int_{t-\alpha}^t\int_{t-\alpha}^t w_r^T g_{\tau,r}^T g_{\tau,\sigma} w_\sigma \, d\sigma\, dr\, d\tau \\
&\qquad = \int_t^\infty\int_{t-\alpha}^t \operatorname{tr}\left[g_{\tau,r} g_{\tau,r}^T\right] dr\, d\tau \\
&\qquad = \alpha\int_t^\infty \operatorname{tr}\left[g_{\tau,t} g_{\tau,t}^T\right] d\tau + \int_t^\infty\int_{t-\alpha}^t \operatorname{tr}\left[g_{\tau,r} g_{\tau,r}^T - g_{\tau,t} g_{\tau,t}^T\right] dr\, d\tau
\end{aligned}$$

Using continuity arguments it can be shown that, given any $\epsilon > 0$, there exists a $\nu > 0$ such that for all $\alpha \in [0, \nu]$ the last term is of order $\alpha\epsilon$. We get

$$\frac{1}{\alpha}\mathcal{E}\,\|z\|_2^2 = Q(G,t) + \mathcal{O}(\alpha^2) + \mathcal{O}(\epsilon)$$

With ϵ being arbitrary, the result follows by taking the limit as α goes to zero. ∎

For systems which admit a state-space realization, this quadratic cost can be expressed in terms of the solution to a Lyapunov equation. Let L_t be the (positive semi-definite bounded) solution to

$$-\dot{L}_t = A_t^T L_t + L_t A_t + C_t^T C_t \tag{7.25}$$

which exists under the assumption that A_t is UES [70]. Then we find the following expression for the quadratic cost operator:

Lemma 7.18 *Suppose that G is given by the stable state-space realization (7.2). Then*

$$Q(G,t) = \operatorname{tr}\left[B_t^T L_t B_t\right]$$

where L_t is the solution to (7.25).

Proof: As in the proof of Lemma 7.8 we can write

$$(G^*Gw)_r = \int_{-\infty}^{r} B_r^T L_r \Phi_{r,\tau} B_\tau w_\tau \, d\tau + \int_r^{\infty} B_r^T \Phi_{\tau,r}^T L_\tau B_\tau w_\tau \, d\tau$$

By taking the same input as in Lemma 7.17 we get

$$Q(G,t) = \lim_{\alpha \downarrow 0} \frac{1}{\alpha} \left(\mathcal{E} \int_{t-\alpha}^{t} \int_{t-\alpha}^{r} w_r^T B_r^T L_r \Phi_{r,\tau} B_\tau w_\tau \, d\tau \, dr \right.$$
$$\left. + \mathcal{E} \int_{t-\alpha}^{t} \int_{r}^{t} w_r^T B_r^T \Phi_{\tau,r}^T L_\tau B_\tau w_\tau \, d\tau \, dr \right)$$

Using similar arguments as in the proof of Lemma 7.17 this can be evaluated as

$$Q(G,t) = \frac{1}{2} \operatorname{tr}\left[B_t^T L_t B_t\right] + \frac{1}{2} \operatorname{tr}\left[B_t^T L_t B_t\right]$$

which completes the proof. ∎

Our aim is to relate the quadratic cost operator to the entropy operator. In fact, we will show that the entropy is an upper bound for the quadratic cost, as in the linear time-invariant case.

Lemma 7.19 *Suppose that G is a causal bounded operator with induced norm less than γ. Then we have that $Q(G,t) \le E(G,\gamma,t)$ for all t.*

Proof: Again we take the same input as in Lemma 7.17. Using the spectral factorization $G^*G = \gamma^2(I - M^*M)$, we get that $Q(G,t)$ equals

$$\lim_{\alpha \downarrow 0} -\frac{\gamma^2}{\alpha} \left(\mathcal{E} \int_{-\infty}^{\infty} w_r^T \int_{-\infty}^{r} m_{r,\tau} w_\tau \, d\tau \, dr + \mathcal{E} \int_{-\infty}^{\infty} w_r^T \int_r^{\infty} m_{\tau,r}^T w_\tau \, d\tau \, dr \right.$$
$$\left. + \mathcal{E} \int_{-\infty}^{\infty} w_r^T \int_r^{\infty} m_{\tau,r}^T \int_{-\infty}^{\tau} m_{\tau,\sigma} w_\sigma d\sigma \, d\tau \, dr \right)$$

The first two terms can be evaluated as

$$-\frac{\gamma^2}{2} \operatorname{tr}[m_{t,t}] - \frac{\gamma^2}{2} \operatorname{tr}\left[m_{t,t}^T\right]$$

which equals the entropy at time t by (7.7). The third term can be evaluated using Lemma 7.17, yielding

$$Q(G,t) = E(G,\gamma,t) - \gamma^2 \int_t^{\infty} \operatorname{tr}\left[m_{\tau,t}^T m_{\tau,t}\right] d\tau \le E(G,\gamma,t) \tag{7.26}$$

which is what we want to show. ∎

In the LTI case the entropy coincides with the $\mathcal{H}_2$ norm if γ tends to infinity. For the time-varying case this is also true. The proof is similar to the proof of Lemma 5.10 for the discrete-time case, though it is very technical, and will be omitted here.

Lemma 7.20 *If G is a strictly causal bounded operator, then*

$$\lim_{\gamma \to \infty} E(G, \gamma, t) = Q(G, t) \qquad \square$$

If G admits a state-space realization, the result is easy to obtain. Since the spectral factor is given with $m_{\tau,t} = -\gamma^{-2} B_\tau^T X_\tau \Phi_{\tau,t} B_t$, where X_t is the solution to (7.17), it follows that

$$-\gamma^2 \int_t^\infty \operatorname{tr}\left[m_{\tau,t}^T m_{\tau,t}\right] d\tau = \mathcal{O}(\gamma^{-2})$$

since A_t is UES. The result follows from the equality in (7.26).

7.4.3 Relationship with risk-sensitive control

We now relate the continuous-time entropy to an LEQG measure for time-varying systems, as was done for discrete-time systems in Section 5.4. We will only consider the finite horizon case (Section 5.5) to illustrate the main ideas. Consider inputs on the interval $[-T, T]$. From (7.1) we see that

$$(Gw)_t = \int_{-T}^t g_{t,\tau} w_\tau \, d\tau \qquad \text{for } t \geq -T$$

We only want to consider the output on the same interval, and since $z_t \neq 0$ for $t > T$ in general, we need a different spectral factorization compared to (7.5). Namely, we consider the factorization $I - \gamma^{-2} G^* G = L^* L$, where

$$\left((I - \gamma^{-2} G^* G) w\right)_t = w_t - \gamma^{-2} \int_t^T g_{\tau,t}^T \int_{-T}^\tau g_{\tau,\sigma} w_\sigma \, d\sigma \, d\tau \qquad (7.27)$$

Note that this enables us to compute $\int_{-T}^T |z_t|^2 \, dt$ as $\int_{-T}^T \langle w_t, (G^* G w)_t \rangle \, dt$. According to (7.7) the entropy can be evaluated as

$$E(G, \gamma, t) = -\frac{\gamma^2}{2} \operatorname{tr}\left[l_{t,t} + l_{t,t}^T\right] \qquad (7.28)$$

We want to relate this entropy to the finite horizon LEQG cost, which is defined in the obvious way as

$$R(G,\theta) = -\frac{2}{\theta}\ln \mathcal{E}\exp\Big(-\frac{\theta}{2}\int_{-T}^{T} |z_t|^2\,dt\Big)$$

where the input is Gaussian white noise on $[-T, T]$. The following theorem, whose proof is found in Appendix B, shows that equality is achieved.

Theorem 7.21 *Suppose that G is a strictly causal operator with norm bound $\|G\| < \gamma$. Then*

$$R(G, -\gamma^{-2}) = \int_{-T}^{T} E(G,\gamma,t)\,dt \qquad \square$$

Connections on an infinite horizon can also be found, similar to those for the discrete-time case in Section 5.4.

In the next section we will prove the main result used for the minimum entropy control problem.

7.5 Minimum entropy control

As in the LTI case, as well as in the discrete-time time-varying case, the minimum entropy control problem is based on writing all the closed-loop systems satisfying a norm bound in a particular form. This form (for the so-called full information case) is the interconnection of an inner system with a free parameter. This parameter has to be chosen so as to minimize the entropy. Since the minimum entropy control problem for the general output feedback case is strongly based on this, the solution to this problem can be derived in a way very similar to the discrete-time time-varying case (Section 6.7). Here we will give the basic result, which can be used directly to solve the full information problem (see Section 6.2 for the discrete-time case).

Before doing so, we should ensure that the cascade of two integral operators is also an integral operator. Suppose that F and H are both causal integral operators with continuous kernels, say

$$(Fw)_t = \hat{f}_t w_t + \int_{-\infty}^{t} f_{t,\tau} w_\tau\,d\tau$$

$$(Hw)_t = \hat{h}_t w_t + \int_{-\infty}^{t} h_{t,\tau} w_\tau\,d\tau \tag{7.29}$$

Then we find the following:

$$\begin{aligned}(FHw)_t &= \hat{f}_t \hat{h}_t w_t + \int_{-\infty}^{t} \hat{f}_t h_{t,\tau} w_\tau \, d\tau + \int_{-\infty}^{t} f_{t,\tau} \hat{h}_\tau w_\tau \, d\tau \\ &\quad + \int_{-\infty}^{t} \int_{-\infty}^{\sigma} f_{t,\sigma} h_{\sigma,\tau} w_\tau \, d\tau \, d\sigma \\ &= \hat{f}_t \hat{h}_t w_t + \int_{-\infty}^{t} \left(\hat{f}_t h_{t,\tau} + f_{t,\tau} \hat{h}_\tau + \int_{\tau}^{t} f_{t,\sigma} h_{\sigma,\tau} d\sigma \right) w_\tau \, d\tau\end{aligned}$$

which is a causal integral operator. Also note that the kernel of this integral operator is continuous since the kernels of F and H are.

An immediate consequence, which we will use to calculate the entropy, is the following: if F and H are both integral operators, given by (7.29) then FH satisfies

$$\begin{aligned}(\hat{fh})_t &= \hat{f}_t \hat{h}_t \\ (fh)_{t,t} &= \hat{f}_t h_{t,t} + f_{t,t} \hat{h}_t \end{aligned} \tag{7.30}$$

For a plant G and a controller K, the closed-loop system is denoted by $\mathcal{F}_\ell(G,K)$, which is a linear fractional transformation of G and K. Without loss of generality we take $\gamma = 1$. If the plant G is given by a state-space realization, then all closed-loop systems satisfying the norm bound $\|\mathcal{F}_\ell(G,K)\| < 1$ are given by $\mathcal{F}_\ell(P,Q)$, where

$$P = \begin{bmatrix} P_{11} & P_{12} \\ P_{21} & P_{22} \end{bmatrix}$$

and Q is any causal operator satisfying $\|Q\| < 1$. The structure of P is such that P_{11} and P_{22} are strictly causal, P_{12} is causal, and P_{21} is a causal operator which has a causal inverse. Furthermore P is inner, i.e. $P^*P = I$. With all these operators being represented as integral operators of the form (7.1), we find the following result for the minimum entropy control problem:

Theorem 7.22 *Suppose that $\mathcal{F}_\ell(P,Q)$ denotes the set of all closed-loop systems, where P is as above, and Q is a causal, bounded contractive integral operator. Then*

(i) $E(\mathcal{F}_\ell(P,Q),1,t)$ is minimized by the unique choice $Q_{\min} \equiv 0$.

(ii) The closed-loop system with minimum entropy equals P_{11}.

Proof: Since

$$\mathcal{F}_\ell(P,Q) = P_{11} + P_{12}Q(I - P_{22}Q)^{-1}P_{21}$$

it is easily checked that, using the assumptions on P, the direct feedthrough of this operator equals $\hat{p}_{11_t} + \hat{p}_{12_t}\hat{q}_t\hat{p}_{21_t}$. This has to be zero in order to have a finite entropy. Since P is inner, it follows that $\hat{p}_{12_t}^T\hat{p}_{12_t} = I$ and $\hat{p}_{21_t}^T\hat{p}_{21_t} = I$. Furthermore P_{21} is square (it is invertible), hence $\hat{q}_t = 0$, which means that Q is strictly causal. To find the entropy of $\mathcal{F}_\ell(P,Q)$, we have to find a spectral factor. Since Q is a contraction, there exists a spectral factorization, say $I - Q^*Q = L^*L$. Using this spectral factor we can write

$$\begin{aligned}
I - &\mathcal{F}_\ell(P,Q)^*\mathcal{F}_\ell(P,Q) \\
&= P_{21}^*\left(I - Q^*P_{22}^*\right)^{-1}(I - Q^*Q)\left(I - P_{22}Q\right)^{-1}P_{21} \\
&= \left[L\left(I - P_{22}Q\right)^{-1}P_{21}\right]^*\left[L\left(I - P_{22}Q\right)^{-1}P_{21}\right] \\
&=: M^*M
\end{aligned}$$

Since $\|P_{22}\| \leq 1$ and $\|Q\| < 1$, the inverse of $I - P_{22}Q$ is well-defined and equals

$$(I - P_{22}Q)^{-1} = \sum_{k=0}^{\infty}(P_{22}Q)^k \tag{7.31}$$

It is easily checked that M is causal and stable, hence it is a spectral factor. For the entropy we are interested in $\hat{m}_t$ and $m_{t,t}$. Without loss of generality we can take the direct feedthrough matrix of L equal to the identity, i.e. $\hat{l}_t = I$. Using (7.30) and (7.31) (note that both P_{22} and Q are strictly causal) we find

$$\begin{aligned}
\hat{m}_t &= \hat{p}_{21_t} \\
m_{t,t} &= l_{t,t}\hat{p}_{21_t} + p_{21_{t,t}}
\end{aligned}$$

Definition 7.2 gives the entropy as

$$\begin{aligned}
E\big(\mathcal{F}_\ell(P,Q),1,t\big) &= \lim_{\alpha\downarrow 0} -\frac{1}{\alpha}\ln\det(\hat{m}_t + \alpha m_{t,t}) \\
&= \lim_{\alpha\downarrow 0} -\frac{1}{\alpha}\ln\det(\hat{p}_{21_t} + \alpha l_{t,t}\hat{p}_{21_t} + \alpha p_{21_{t,t}}) \\
&= \lim_{\alpha\downarrow 0} -\frac{1}{\alpha}\ln\det\big((I + \alpha l_{t,t})(\hat{p}_{21_t} + \alpha p_{21_{t,t}}) + \mathcal{O}(\alpha^2)\big)
\end{aligned}$$

Dropping the higher order terms results in

$$\begin{aligned}
&E\big(\mathcal{F}_\ell(P,Q),1,t\big) \\
&\qquad = \lim_{\alpha\downarrow 0} -\frac{1}{\alpha}\ln\det\big((I+\alpha l_{t,t})(\hat{p}_{21_t}+\alpha p_{21_{t,t}})\big) \\
&\qquad = \lim_{\alpha\downarrow 0} -\frac{1}{\alpha}\ln\det(\hat{p}_{21_t}+\alpha p_{21_{t,t}}) + \lim_{\alpha\downarrow 0} -\frac{1}{\alpha}\ln\det(I+\alpha l_{t,t}) \\
&\qquad = E(P_{11},1,t)+E(Q,1,t)
\end{aligned}$$

where the last step follows from the definition of L, and the fact that $I - P_{11}^*P_{11} = P_{21}^*P_{21}$ since P is inner. The first term is fixed, and the second term is minimized uniquely by $Q \equiv 0$ by Lemma 7.5. ∎

A Proof of Theorem 6.5

The proof of Theorem 6.5 mimics the proof of a similar result in the discrete-time $\mathcal{H}_\infty$ problem for LTI systems, found in [75]. In the proof we use an initial condition, and so will make use of single-infinite sequences. Without loss of generality, we can choose the initial time $k_0 = 0$. By letting the initial time go to minus infinity, we will get the general result.

The operator $\boldsymbol{G}$ is now a single-infinite operator defined by $G_{i,j}$ for $i, j \geq 0$, mapping ℓ_{2+}^m to ℓ_{2+}^p. Notice that now the forward shift operator is only right-invertible ($\boldsymbol{Z}\boldsymbol{Z}^* = \boldsymbol{I}$), and that $\boldsymbol{Z}^*\boldsymbol{Z} = \boldsymbol{I} - \Delta_0$, where

$$\Delta_k \boldsymbol{x} := \begin{bmatrix} \cdots & 0 & x_k^T & 0 & \cdots \end{bmatrix}^T$$

For a system $\boldsymbol{G}$, given by the state-space realization (2.1), we will shortly outline the effect of incorporating an initial state. Suppose that the initial state is a vector $x_0 \in \mathbb{R}^n$. The sequence $\boldsymbol{x}_0$ denotes the initial state sequence $\Delta_0 \boldsymbol{x} = (x_0^T \; 0^T \; 0^T \; \ldots)^T \in \ell_{2+}^n$. Thus, we have that the output $\boldsymbol{z} = \boldsymbol{z}_{x_0} + \boldsymbol{z}_w$ to (2.1) consists of the autonomous output given by $\boldsymbol{z}_x := \boldsymbol{z}|_{\boldsymbol{w} \equiv \boldsymbol{0}} = \boldsymbol{C}\boldsymbol{\Phi}\boldsymbol{x}_0$, and the response to the input $\boldsymbol{w}$: $\boldsymbol{z}_w := \boldsymbol{z}|_{x_0 = 0} = \boldsymbol{G}\boldsymbol{w}$.

After this introduction we will give the proof of Theorem 6.5, consisting of several claims. To make the computations less complicated we will make the additional assumption that

$$\boldsymbol{D}_{12}^* \begin{bmatrix} \boldsymbol{C}_1 & \boldsymbol{D}_{11} \end{bmatrix} = \begin{bmatrix} \boldsymbol{0} & \boldsymbol{0} \end{bmatrix} \tag{A.1}$$

and the general statement can be proven by applying a preliminary feedback (see Remark A.1). Notice that assumption *A(5)* in the problem formulation (Section 6.1) is now simplified to $(\boldsymbol{C}_1, \boldsymbol{A})$ being uniformly detectable.

We consider the problem of finding $\boldsymbol{w}$ and $\boldsymbol{u}$ solving

$$J(x_0) := \sup_{\boldsymbol{w}\in \ell_{2+}^{m_1}} \inf_{\boldsymbol{u}} \left\{ \|\boldsymbol{z}\|_2^2 - \|\boldsymbol{w}\|_2^2 \,\middle|\, \boldsymbol{u} \in \ell_{2+}^{m} \text{ such that } \boldsymbol{x} \in \ell_{2+}^{n} \right\} \tag{A.2}$$

where x_0 is the initial condition. First we will take a fixed $\boldsymbol{w} \in \ell_{2+}^{m_1}$, and consider the problem of infimizing over $\boldsymbol{u}$. Let $\boldsymbol{L} \geq \boldsymbol{0}$ be the memoryless solution of

$$\boldsymbol{L} = \boldsymbol{A}^*\boldsymbol{Z}\boldsymbol{L}\boldsymbol{Z}^*\boldsymbol{A} + \boldsymbol{C}_1^*\boldsymbol{C}_1 - \boldsymbol{A}^*\boldsymbol{Z}\boldsymbol{L}\boldsymbol{Z}^*\boldsymbol{B}_2\boldsymbol{N}^{-1}\boldsymbol{B}_2^*\boldsymbol{Z}\boldsymbol{L}\boldsymbol{Z}^*\boldsymbol{A} \tag{A.3}$$

for which

$$\tilde{\boldsymbol{A}} := \boldsymbol{A} - \boldsymbol{B}_2\boldsymbol{N}^{-1}\boldsymbol{B}_2^*\boldsymbol{Z}\boldsymbol{L}\boldsymbol{Z}^*\boldsymbol{A}$$

is UES, where $\boldsymbol{N} := \boldsymbol{D}_{12}^*\boldsymbol{D}_{12} + \boldsymbol{B}_2^*\boldsymbol{Z}\boldsymbol{L}\boldsymbol{Z}^*\boldsymbol{B}_2 > \boldsymbol{0}$. It is shown in [27] that $\boldsymbol{L}$ exists under the assumptions made. We can write $\tilde{\boldsymbol{A}} = \boldsymbol{U}^*\boldsymbol{A}$, where

$$\boldsymbol{U} = \boldsymbol{I} - \boldsymbol{Z}\boldsymbol{L}\boldsymbol{Z}^*\boldsymbol{B}_2\boldsymbol{N}^{-1}\boldsymbol{B}_2^*$$

Define

$$\tilde{\boldsymbol{u}} = -\boldsymbol{N}^{-1}\boldsymbol{B}_2^*\boldsymbol{Z}\boldsymbol{L}\boldsymbol{Z}^*\boldsymbol{A}\boldsymbol{x} - \boldsymbol{N}^{-1}\boldsymbol{B}_2^*(\boldsymbol{I} - \boldsymbol{Z}\boldsymbol{A}^*\boldsymbol{U})^{-1}\boldsymbol{Z}(\boldsymbol{L}\boldsymbol{Z}^*\boldsymbol{B}_1 + \boldsymbol{C}_1^*\boldsymbol{D}_{11})\boldsymbol{w}$$

Applying this feedback to the original plant results in

$$\Sigma_{\tilde{G}} = \begin{cases} \boldsymbol{Z}\boldsymbol{x} &= \tilde{\boldsymbol{A}}\boldsymbol{x} + \tilde{\boldsymbol{B}}\boldsymbol{w} \\ \boldsymbol{z} &= \tilde{\boldsymbol{C}}\boldsymbol{x} + \tilde{\boldsymbol{D}}\boldsymbol{w} \end{cases} \tag{A.4}$$

where

$$\begin{aligned}
\tilde{\boldsymbol{B}} &:= \boldsymbol{B}_1 - \boldsymbol{B}_2\boldsymbol{N}^{-1}\boldsymbol{B}_2^*(\boldsymbol{I} - \boldsymbol{Z}\boldsymbol{A}^*\boldsymbol{U})^{-1}\boldsymbol{Z}(\boldsymbol{L}\boldsymbol{Z}^*\boldsymbol{B}_1 + \boldsymbol{C}_1^*\boldsymbol{D}_{11}) \\
\tilde{\boldsymbol{C}} &:= \boldsymbol{C}_1 - \boldsymbol{D}_{12}\boldsymbol{N}^{-1}\boldsymbol{B}_2^*\boldsymbol{Z}\boldsymbol{L}\boldsymbol{Z}^*\boldsymbol{A} \\
\tilde{\boldsymbol{D}} &:= \boldsymbol{D}_{11} - \boldsymbol{D}_{12}\boldsymbol{N}^{-1}\boldsymbol{B}_2^*(\boldsymbol{I} - \boldsymbol{Z}\boldsymbol{A}^*\boldsymbol{U})^{-1}\boldsymbol{Z}(\boldsymbol{L}\boldsymbol{Z}^*\boldsymbol{B}_1 + \boldsymbol{C}_1^*\boldsymbol{D}_{11})
\end{aligned}$$

In the sequel we will denote by $\tilde{\boldsymbol{x}}$ and $\tilde{\boldsymbol{z}}$ the state and the output of $\tilde{\boldsymbol{G}}$ given by (A.4), i.e. the system $\boldsymbol{G}$ with $\tilde{\boldsymbol{u}}$ as control input. Since $\tilde{\boldsymbol{A}}$ is UES it can easily be shown that, for $\boldsymbol{w} \in \ell_{2+}^{m_1}$, we have $\tilde{\boldsymbol{u}} \in \ell_{2+}^{m_2}$. Furthermore we define

$$\eta := -\boldsymbol{Z}\boldsymbol{L}\boldsymbol{Z}^*\boldsymbol{U}^*\boldsymbol{A}\tilde{\boldsymbol{x}} - \boldsymbol{U}(\boldsymbol{I} - \boldsymbol{Z}\boldsymbol{A}^*\boldsymbol{U})^{-1}\boldsymbol{Z}(\boldsymbol{L}\boldsymbol{Z}^*\boldsymbol{B}_1 + \boldsymbol{C}_1^*\boldsymbol{D}_{11})\boldsymbol{w}$$

Claim 1 *η satisfies*

$$\begin{aligned} Z^*\eta &= A^*\eta - C_1^*C_1\tilde{x} - C_1^*D_{11}w + \Delta_0 f, \qquad \lim_{k\to\infty} \Delta_k\eta = 0 \\ f &:= L\tilde{x} + (I - A^*UZ)^{-1}(LZ^*B_1 + C_1^*D_{11})w \end{aligned} \tag{A.5}$$

Proof: First note that $Z^*ZLZ^* = (I - \Delta_0)LZ^* = LZ^*$. Using this, (A.3) and (A.4), we find

$$\begin{aligned} &(Z^* - A^*)\eta \\ &\quad = -(Z^* - A^*)ZLZ^*U^*A\tilde{x} \\ &\qquad - (Z^* - A^*)U(I - ZA^*U)^{-1}Z(LZ^*B_1 + C_1^*D_{11})w \\ &\quad = -Z^*ZLZ^*U^*A\tilde{x} + A^*ZLZ^*U^*A\tilde{x} \\ &\qquad - (Z^*UZ - A^*UZ)(I - A^*UZ)^{-1}(LZ^*B_1 + C_1^*D_{11})w \\ &\quad = -LZ^*U^*A\tilde{x} + L\tilde{x} - C_1^*C_1\tilde{x} - (LZ^*B_1 + C_1^*D_{11})w \\ &\qquad - (Z^*UZ - I)(I - A^*UZ)^{-1}(LZ^*B_1 + C_1^*D_{11})w \\ &\quad = -LZ^*U^*A\tilde{x} + LZ^*Z\tilde{x} + L\Delta_0\tilde{x} - C_1^*C_1\tilde{x} - (LZ^*B_1 + C_1^*D_{11})w \\ &\qquad + (LZ^*B_2N^{-1}B_2^*Z + \Delta_0)(I - A^*UZ)^{-1}(LZ^*B_1 + C_1^*D_{11})w \\ &\quad = -C_1^*C_1\tilde{x} - C_1^*D_{11}w + LZ^*(Z\tilde{x} - \tilde{A}\tilde{x} - \tilde{B}w) \\ &\qquad + \Delta_0\left(L\tilde{x} + (I - A^*UZ)^{-1}(LZ^*B_1 + C_1^*D_{11})w\right) \\ &\quad = -C_1^*C_1\tilde{x} - C_1^*D_{11}w + \Delta_0 f \end{aligned}$$

Furthermore it is easily seen that $\Delta_k\eta$ is a function of $\Delta_k\tilde{x}_k$ and $P_{k-1}^{\perp}w$, and since $\tilde{A}$ is UES and $w \in \ell_{2+}^{m_1}$, the end condition on η follows immediately.

Claim 2 $B_2^*\eta = D_{12}^*D_{12}\tilde{u}$

Proof: By definition of N we have

$$\left(D_{12}^*D_{12} + B_2^*ZLZ^*B_2\right)N^{-1}B_2^* = B_2^*$$

and hence

$$\begin{aligned} D_{12}^*D_{12}N^{-1}B_2^* &= B_2^* - B_2^*ZLZ^*B_2N^{-1}B_2^* \\ &= B_2^*U \end{aligned}$$

Now, since

$$UZLZ^* = ZLZ^*U^*$$

the result follows easily from the definitions of η and $\tilde{u}$.

Claim 3 *Given x_0 and $\boldsymbol{w} \in \ell_{2+}^{m_1}$, then*

$$\tilde{\boldsymbol{u}} = \arg\inf_{\boldsymbol{u}} \left\{ \|\boldsymbol{z}\|_2^2 \;\middle|\; \boldsymbol{u} \in \ell_{2+}^{m_2} \quad \text{such that } \boldsymbol{x} \in \ell_{2+}^{n} \right\}$$

Proof: A little algebra yields

$$\begin{aligned}
\|\tilde{\boldsymbol{z}}\|_2^2 - \|\boldsymbol{z}\|_2^2 &= \|\boldsymbol{C}_1\tilde{\boldsymbol{x}} + \boldsymbol{D}_{11}\boldsymbol{w} + \boldsymbol{D}_{12}\tilde{\boldsymbol{u}}\|_2^2 - \|\boldsymbol{C}_1\boldsymbol{x} + \boldsymbol{D}_{11}\boldsymbol{w} + \boldsymbol{D}_{12}\boldsymbol{u}\|_2^2 \\
&\quad - 2\langle \eta, \boldsymbol{Z}\tilde{\boldsymbol{x}}\rangle + 2\langle \boldsymbol{Z}^*\eta, \tilde{\boldsymbol{x}}\rangle + 2\langle \eta, \boldsymbol{Z}\boldsymbol{x}\rangle - 2\langle \boldsymbol{Z}^*\eta, \boldsymbol{x}\rangle \\
&= -\|\boldsymbol{C}_1(\boldsymbol{x} - \tilde{\boldsymbol{x}})\|_2^2 + 2\|\boldsymbol{C}_1\tilde{\boldsymbol{x}}\|_2^2 - 2\langle \boldsymbol{C}_1\boldsymbol{x}, \boldsymbol{C}_1\tilde{\boldsymbol{x}}\rangle \\
&\quad + \|\boldsymbol{D}_{12}\tilde{\boldsymbol{u}}\|_2^2 - \|\boldsymbol{D}_{12}\boldsymbol{u}\|_2^2 + 2\langle \boldsymbol{C}_1\tilde{\boldsymbol{x}}, \boldsymbol{D}_{11}\boldsymbol{w}\rangle \\
&\quad - 2\langle \boldsymbol{C}_1\boldsymbol{x}, \boldsymbol{D}_{11}\boldsymbol{w}\rangle - 2\langle \eta, \boldsymbol{U}^*\boldsymbol{A}\tilde{\boldsymbol{x}} + \boldsymbol{B}_2\tilde{\boldsymbol{u}} + \boldsymbol{B}_1\boldsymbol{w}\rangle \\
&\quad + 2\langle \boldsymbol{A}^*\eta - \boldsymbol{C}_1^*\boldsymbol{C}_1\tilde{\boldsymbol{x}} - \boldsymbol{C}_1^*\boldsymbol{D}_{11}\boldsymbol{w} + \Delta_0\boldsymbol{f}, \tilde{\boldsymbol{x}}\rangle \\
&\quad + 2\langle \eta, \boldsymbol{A}\boldsymbol{x} + \boldsymbol{B}_2\boldsymbol{u} + \boldsymbol{B}_1\boldsymbol{w}\rangle \\
&\quad - 2\langle \boldsymbol{A}^*\eta - \boldsymbol{C}_1^*\boldsymbol{C}_1\tilde{\boldsymbol{x}} - \boldsymbol{C}_1^*\boldsymbol{D}_{11}\boldsymbol{w} + \Delta_0\boldsymbol{f}, \boldsymbol{x}\rangle \\
&= -\|\boldsymbol{C}_1(\boldsymbol{x} - \tilde{\boldsymbol{x}})\|_2^2 + \langle \boldsymbol{D}_{12}^*\boldsymbol{D}_{12}\tilde{\boldsymbol{u}} - 2\boldsymbol{B}_2^*\eta, \tilde{\boldsymbol{u}}\rangle \\
&\quad - \langle \boldsymbol{D}_{12}^*\boldsymbol{D}_{12}\boldsymbol{u} - 2\boldsymbol{B}_2^*\eta, \boldsymbol{u}\rangle \\
&= -\|\boldsymbol{C}_1(\boldsymbol{x} - \tilde{\boldsymbol{x}})\|_2^2 - \|\boldsymbol{D}_{12}(\boldsymbol{u} - \tilde{\boldsymbol{u}})\|_2^2
\end{aligned}$$

which is non-positive. Since $\boldsymbol{D}_{12}^*\boldsymbol{D}_{12} > \boldsymbol{0}$ it follows also that $\tilde{\boldsymbol{u}}$ is the unique minimizing control.

Claim 4 *Given x_0, $\sup_{\boldsymbol{w}\in\ell_{2+}^{m_1}} \left\{ \|\tilde{\boldsymbol{z}}\|_2^2 - \|\boldsymbol{w}\|_2^2 \right\}$ is bounded above and below:*

$$x_0^T L_0 x_0 \le \sup_{\boldsymbol{w}\in\ell_{2+}^{m_1}} \left\{ \|\tilde{\boldsymbol{z}}\|_2^2 - \|\boldsymbol{w}\|_2^2 \right\} < \infty$$

where we write $\boldsymbol{L}$ as $\boldsymbol{L} = \operatorname{diag}\{L_0, L_1, \ldots\}$.

Proof: By setting $\boldsymbol{w} \equiv \boldsymbol{0}$, we have that $\sup_{\boldsymbol{w}\in\ell_{2+}^{m_1}} \left\{ \|\tilde{\boldsymbol{z}}\|_2^2 - \|\boldsymbol{w}\|_2^2 \right\} \ge \|\tilde{\boldsymbol{z}}_x\|_2^2$, and by noting that the closed-loop matrices satisfy $\boldsymbol{L} = \tilde{\boldsymbol{A}}^*\boldsymbol{Z}\boldsymbol{L}\boldsymbol{Z}^*\tilde{\boldsymbol{A}} + \tilde{\boldsymbol{C}}^*\tilde{\boldsymbol{C}}$, we get

$$\begin{aligned}
\|\tilde{\boldsymbol{z}}_x\|_2^2 &= \langle \tilde{\boldsymbol{z}}_x, \tilde{\boldsymbol{z}}_x\rangle + \langle \tilde{\boldsymbol{x}}, \boldsymbol{L}\tilde{\boldsymbol{x}}\rangle - \langle \tilde{\boldsymbol{x}}, \boldsymbol{L}\tilde{\boldsymbol{x}}\rangle \\
&= \langle \tilde{\boldsymbol{C}}\tilde{\boldsymbol{x}}, \tilde{\boldsymbol{C}}\tilde{\boldsymbol{x}}\rangle + \langle \boldsymbol{Z}\tilde{\boldsymbol{x}}, \boldsymbol{Z}\boldsymbol{L}\tilde{\boldsymbol{x}}\rangle - \langle \tilde{\boldsymbol{x}}, \boldsymbol{L}\tilde{\boldsymbol{x}}\rangle + \langle \tilde{\boldsymbol{x}}, \Delta_0\boldsymbol{L}\tilde{\boldsymbol{x}}\rangle \\
&= \langle \tilde{\boldsymbol{x}}, \tilde{\boldsymbol{C}}^*\tilde{\boldsymbol{C}}\tilde{\boldsymbol{x}}\rangle + \langle \tilde{\boldsymbol{A}}\tilde{\boldsymbol{x}}, \boldsymbol{Z}\boldsymbol{L}\boldsymbol{Z}^*\tilde{\boldsymbol{A}}\tilde{\boldsymbol{x}}\rangle + \langle \tilde{\boldsymbol{A}}\tilde{\boldsymbol{x}}, \boldsymbol{Z}\boldsymbol{L}\Delta_0\tilde{\boldsymbol{x}}\rangle - \langle \tilde{\boldsymbol{x}}, \boldsymbol{L}\tilde{\boldsymbol{x}}\rangle \\
&\quad + {x_0}^T L_0 x_0 \\
&= {x_0}^T L_0 x_0
\end{aligned}$$

where we have used the fact that $\boldsymbol{ZL}\Delta_0 = \mathbf{0}$. Now, assume without loss of generality that $\|\boldsymbol{w}\|_2 \neq 0$. From Claim 3 and the assumption in Theorem 6.5 we know there exists a $\delta > 0$ such that $\|\tilde{\boldsymbol{G}}\| \leq \sqrt{(1-\delta^2)}$. Hence

$$\begin{aligned}\|\tilde{\boldsymbol{z}}\|_2^2 - \|\boldsymbol{w}\|_2^2 &= \|\tilde{\boldsymbol{z}}_x + \tilde{\boldsymbol{z}}_w\|_2^2 - \|\boldsymbol{w}\|_2^2 \\ &\leq \left(\left(\|\tilde{\boldsymbol{z}}_x\|_2 / \|\boldsymbol{w}\|_2 + \|\tilde{\boldsymbol{G}}\boldsymbol{w}\|_2 / \|\boldsymbol{w}\|_2\right)^2 - 1\right)\|\boldsymbol{w}\|_2^2 \\ &\leq \left(\left(\sqrt{x_0^T L_0 x_0} / \|\boldsymbol{w}\|_2 + \sqrt{(1-\delta^2)}\right)^2 - 1\right)\|\boldsymbol{w}\|_2^2\end{aligned}$$

which is bounded above (notice that the last expression is negative when $\|\boldsymbol{w}\|_2^2 \to \infty$).

Claim 5 *For all x_0, there exists a unique $\boldsymbol{w}^o \in \ell_{2+}^{m_1}$ such that the supremum in (A.2) is achieved.*

Proof: Consider a maximizing sequence $\{\boldsymbol{w}_k\}$ of $\sup_{\boldsymbol{w}\in\ell_{2+}^{m_1}} \left\{\|\tilde{\boldsymbol{z}}\|_2^2 - \|\boldsymbol{w}\|_2^2\right\}$. For $\boldsymbol{w}_\alpha$, $\boldsymbol{w}_\beta \in \ell_{2+}^{m_1}$ applied to $\tilde{\boldsymbol{G}}$, denote the respective outputs by $\tilde{\boldsymbol{z}}_{w_\alpha}$ and $\tilde{\boldsymbol{z}}_{w_\beta}$. Let $\boldsymbol{w}_\kappa := (\boldsymbol{w}_\alpha + \boldsymbol{w}_\beta)/2$ and $\tilde{\boldsymbol{z}}_{w_\kappa}$ the corresponding output. With a little algebra, we can show that:

$$\begin{aligned}&\|\tilde{\boldsymbol{G}}(\boldsymbol{w}_\alpha - \boldsymbol{w}_\beta)\|_2^2 - \|\boldsymbol{w}_\alpha - \boldsymbol{w}_\beta\|_2^2 \\ &\quad = \|\tilde{\boldsymbol{z}}_x + \tilde{\boldsymbol{G}}\boldsymbol{w}_\alpha - (\tilde{\boldsymbol{z}}_x + \tilde{\boldsymbol{G}}\boldsymbol{w}_\beta)\|_2^2 - \|\boldsymbol{w}_\alpha - \boldsymbol{w}_\beta\|_2^2 \qquad \text{(A.6)} \\ &\quad = 2\left(\|\tilde{\boldsymbol{z}}_x + \tilde{\boldsymbol{z}}_{w_\alpha}\|_2^2 - \|\boldsymbol{w}_\alpha\|_2^2\right) + 2\left(\|\tilde{\boldsymbol{z}}_x + \tilde{\boldsymbol{z}}_{w_\beta}\|_2^2 - \|\boldsymbol{w}_\beta\|_2^2\right) \\ &\qquad - 4\left(\|\tilde{\boldsymbol{z}}_x + \tilde{\boldsymbol{z}}_{w_\kappa}\|_2^2 - \|\boldsymbol{w}_\kappa\|_2^2\right) \\ &\quad \geq 2\left(\|\tilde{\boldsymbol{z}}_x + \tilde{\boldsymbol{z}}_{w_\alpha}\|_2^2 - \|\boldsymbol{w}_\alpha\|_2^2\right) + 2\left(\|\tilde{\boldsymbol{z}}_x + \tilde{\boldsymbol{z}}_{w_\beta}\|_2^2 - \|\boldsymbol{w}_\beta\|_2^2\right) \\ &\qquad - 4 \sup_{\boldsymbol{w}\in\ell_{2+}^{m_1}} \left\{\|\tilde{\boldsymbol{z}}_x + \tilde{\boldsymbol{z}}_w\|_2^2 - \|\boldsymbol{w}\|_2^2\right\}\end{aligned}$$

Since the sequence $\{\boldsymbol{w}_\alpha\}$ is maximizing, the right hand side of this equation will go to 0. Thus, for all $\epsilon > 0$ there exists an N such that for all α, $\beta > N$

$$\|\boldsymbol{w}_\alpha - \boldsymbol{w}_\beta\|_2^2 < \epsilon + \|\tilde{\boldsymbol{G}}(\boldsymbol{w}_\alpha - \boldsymbol{w}_\beta)\|_2^2 \leq \epsilon + \|\tilde{\boldsymbol{G}}\|^2 \|\boldsymbol{w}_\alpha - \boldsymbol{w}_\beta\|_2^2$$

This implies that $\{\boldsymbol{w}_k\}$ is a Cauchy sequence in $\ell_{2+}^{m_1}$, from which we can conclude that $\lim_{k\to\infty} \boldsymbol{w}_k = \boldsymbol{w}^o$.

Suppose that $\boldsymbol{w}^o$ is not unique, say $\bar{\boldsymbol{w}}$ is also maximizing. We know that $\|\tilde{\boldsymbol{G}}\| \leq \sqrt{(1-\delta^2)}$, hence

$$-\|\boldsymbol{w}\|_2^2 \leq \|\tilde{\boldsymbol{G}}\boldsymbol{w}\|_2^2 - \|\boldsymbol{w}\|_2^2 \leq -\delta^2 \|\boldsymbol{w}\|_2^2$$

from which it follows that $\sqrt{\|\boldsymbol{w}\|_2^2 - \|\tilde{\boldsymbol{G}}\boldsymbol{w}\|_2^2}$ is a norm on ℓ_{2+}, equivalent to the usual ℓ_{2+} norm. As in the derivation (A.6), with $\boldsymbol{w}_\alpha$ and $\boldsymbol{w}_\beta$ replaced by $\boldsymbol{w}^o$ and $\bar{\boldsymbol{w}}$, and $\boldsymbol{w}_\kappa$ by $(\boldsymbol{w}^o+\bar{\boldsymbol{w}})/2$, we see that the right-hand side equals 0. And hence the left-hand side (which is always less than or equal to 0) equals 0, from which it follows that $\bar{\boldsymbol{w}} = \boldsymbol{w}^o$.

In the sequel the superscript $(\cdot)^o$ denotes optimality, i.e. the system with the worst-case disturbance $\boldsymbol{w}^o$.

Claim 6 *Given x_0, the unique optimal $\boldsymbol{w}^o$ satisfies*

$$(\boldsymbol{I} - \boldsymbol{D}_{11}^*\boldsymbol{D}_{11})w = -\boldsymbol{B}_1^*\eta + \boldsymbol{D}_{11}^*\boldsymbol{C}_1\tilde{\boldsymbol{x}} \tag{A.7}$$

Proof: Denote by $\boldsymbol{w}^o := \arg \sup_{\boldsymbol{w}\in\ell_{2+}^{m_1}} \left\{ \|\tilde{z}\|_2^2 - \|\boldsymbol{w}\|_2^2 \right\}$ and $\tilde{\boldsymbol{x}}^o$, η^o, $\tilde{\boldsymbol{u}}^o$ and $\tilde{z}^o$ the corresponding states and outputs due to this input. Moreover let $\bar{\boldsymbol{w}}$ be defined by

$$\bar{\boldsymbol{w}} = -\boldsymbol{B}_1^*\eta^o + \boldsymbol{D}_{11}^*\boldsymbol{C}_1\tilde{\boldsymbol{x}}^o + \boldsymbol{D}_{11}^*\boldsymbol{D}_{11}\boldsymbol{w}^o$$

Recalling that

$$\boldsymbol{D}_{12}^*\begin{bmatrix} \boldsymbol{C}_1 & \boldsymbol{D}_{11} \end{bmatrix} = \begin{bmatrix} \boldsymbol{0} & \boldsymbol{0} \end{bmatrix}$$

and

$$\boldsymbol{B}_2^*\eta = \boldsymbol{D}_{12}^*\boldsymbol{D}_{12}\tilde{\boldsymbol{u}}$$

we can show that

$$\begin{aligned}
&\|\tilde{z}^o - \bar{z}\|_2^2 - \|\tilde{z}^o\|_2^2 - \|\bar{z}\|_2^2 \\
&\quad = 2\langle \boldsymbol{Z}\bar{\boldsymbol{x}}, \eta^o\rangle - 2\langle \bar{\boldsymbol{x}}, \boldsymbol{Z}^*\eta^o\rangle - 2\langle \bar{z}, \tilde{z}^o\rangle \\
&\quad = 2\langle \boldsymbol{A}\bar{\boldsymbol{x}} + \boldsymbol{B}_1\bar{\boldsymbol{w}} + \boldsymbol{B}_2\bar{\boldsymbol{u}}, \eta^o\rangle \\
&\qquad - 2\langle \bar{\boldsymbol{x}}, \boldsymbol{A}^*\eta^o - \boldsymbol{C}_1^*\boldsymbol{C}_1\tilde{\boldsymbol{x}}^o - \boldsymbol{C}_1^*\boldsymbol{D}_{11}\boldsymbol{w}^o + \Delta_0\boldsymbol{f}^o\rangle \\
&\qquad - 2\langle \boldsymbol{C}_1\bar{\boldsymbol{x}} + \boldsymbol{D}_{11}\bar{\boldsymbol{w}} + \boldsymbol{D}_{12}\bar{\boldsymbol{u}}, \boldsymbol{C}_1\tilde{\boldsymbol{x}}^o + \boldsymbol{D}_{11}\boldsymbol{w}^o + \boldsymbol{D}_{12}\tilde{\boldsymbol{u}}\rangle \\
&\quad = -2\langle \bar{\boldsymbol{w}}, -\boldsymbol{B}_1^*\eta^o + \boldsymbol{D}_{11}^*\boldsymbol{C}_1\tilde{\boldsymbol{x}}^o + \boldsymbol{D}_{11}^*\boldsymbol{D}_{11}\boldsymbol{w}^o\rangle - 2\langle \bar{\boldsymbol{x}}, \Delta_0\boldsymbol{f}^o\rangle \\
&\quad = -2\|\bar{\boldsymbol{w}}\|_2^2 - 2\langle \bar{\boldsymbol{x}}, \Delta_0\boldsymbol{f}^o\rangle
\end{aligned} \tag{A.8}$$

where we have used (A.5) in the second equality, and Claim 2 in the third. Similarly

$$\begin{aligned}
\|\tilde{\boldsymbol{z}}^o\|_2^2 &= \langle \tilde{\boldsymbol{z}}^o, \tilde{\boldsymbol{z}}^o \rangle - \langle \eta^o, \boldsymbol{Z}\tilde{\boldsymbol{x}}^o \rangle + \langle \boldsymbol{Z}^*\eta_*, \boldsymbol{x}^o \rangle \\
&= \langle \boldsymbol{C}_1\tilde{\boldsymbol{x}}^o + \boldsymbol{D}_{11}\boldsymbol{w}^o + \boldsymbol{D}_{12}\tilde{\boldsymbol{u}}^o, \boldsymbol{C}_1\tilde{\boldsymbol{x}}^o + \boldsymbol{D}_{11}\boldsymbol{w}^o + \boldsymbol{D}_{12}\tilde{\boldsymbol{u}}^o \rangle \\
&\quad - \langle \eta^o, \boldsymbol{A}\tilde{\boldsymbol{x}}^o + \boldsymbol{B}_1\boldsymbol{w}^o + \boldsymbol{B}_2\tilde{\boldsymbol{u}}^o \rangle \\
&\quad + \langle \boldsymbol{A}^*\eta^o - \boldsymbol{C}_1^*\boldsymbol{C}_1\tilde{\boldsymbol{x}}^o - \boldsymbol{C}_1^*\boldsymbol{D}_{11}\boldsymbol{w}^o + \Delta_0 f^o, \tilde{\boldsymbol{x}}^o \rangle \\
&= \langle -\boldsymbol{B}_1^*\eta^o + \boldsymbol{D}_{11}^*\boldsymbol{C}_1\tilde{\boldsymbol{x}}^o + \boldsymbol{D}_{11}^*\boldsymbol{D}_{11}\boldsymbol{w}^o, \boldsymbol{w}^o \rangle \\
&\quad + \langle \tilde{\boldsymbol{x}}^o, \Delta_0 \boldsymbol{f}^o \rangle \\
&= \langle \tilde{\boldsymbol{x}}^o, \Delta_0 \boldsymbol{f}^o \rangle
\end{aligned}$$

and hence

$$\|\tilde{\boldsymbol{z}}^o\|_2^2 - \langle \bar{\boldsymbol{w}}, \boldsymbol{w}^o \rangle = \langle \tilde{\boldsymbol{x}}^o, \Delta_0 \boldsymbol{f}^o \rangle \tag{A.9}$$

from the definition of $\bar{\boldsymbol{w}}$.

Combining the expressions (A.8) and (A.9) results in

$$\left(\|\bar{\boldsymbol{z}}\|_2^2 - \|\bar{\boldsymbol{w}}\|_2^2\right) - \left(\|\tilde{\boldsymbol{z}}^o\|_2^2 - \|\boldsymbol{w}^o\|_2^2\right) = \|\tilde{\boldsymbol{z}}^o - \bar{\boldsymbol{z}}\|_2^2 + \|\boldsymbol{w}^o - \bar{\boldsymbol{w}}\|_2^2 \geq 0$$

where we used that $\bar{x}_0 = \tilde{x}_0^o = x_0$. Since $\boldsymbol{w}^o$ is maximizing, it follows that $\bar{\boldsymbol{w}} = \boldsymbol{w}^o \in \ell_{2+}^{m_1}$.

Claim 7 *There exist bounded, memoryless operators $\boldsymbol{H}_1$ and $\boldsymbol{H}_2$ such that $\boldsymbol{w}^o = \boldsymbol{H}_1\tilde{\boldsymbol{x}}^o$ and $\eta^o = \boldsymbol{H}_2\tilde{\boldsymbol{x}}^o$.*

Proof: From the definitions of the closed-loop matrices it is straightforward to write equations (A.5) and (A.7) as

$$\begin{aligned}
\boldsymbol{Z}^*\eta^o &= \tilde{\boldsymbol{A}}^*\eta^o - \tilde{\boldsymbol{C}}^*\tilde{\boldsymbol{C}}\tilde{\boldsymbol{x}}^o - \tilde{\boldsymbol{C}}^*\tilde{\boldsymbol{D}}\boldsymbol{w}^o + \Delta_0\boldsymbol{f}^o, \quad \lim_{k\to\infty} \Delta_k\eta^o = \boldsymbol{0} \\
(\boldsymbol{I} - \tilde{\boldsymbol{D}}^*\tilde{\boldsymbol{D}})\boldsymbol{w}^o &= -\tilde{\boldsymbol{B}}^*\eta^o + \tilde{\boldsymbol{D}}^*\tilde{\boldsymbol{C}}\tilde{\boldsymbol{x}}^o
\end{aligned}$$

Furthermore recall that, for a given x_0, the state $\tilde{\boldsymbol{x}}$ can be written as

$$\tilde{\boldsymbol{x}} = \tilde{\Phi}\boldsymbol{x}_0 + \tilde{\Phi}\boldsymbol{Z}^*\tilde{\boldsymbol{B}}\boldsymbol{w}$$

where $\tilde{\Phi} = (\boldsymbol{I} - \boldsymbol{Z}^*\tilde{\boldsymbol{A}})^{-1}$.

Using this equation, the co-state η^o and input $\boldsymbol{w}^o$ can be written as

$$\begin{aligned}
\eta^o &= -\boldsymbol{Z}\tilde{\Phi}^*\tilde{\boldsymbol{C}}^*(\tilde{\boldsymbol{C}}\tilde{\boldsymbol{x}}^o + \tilde{\boldsymbol{D}}\boldsymbol{w}^o) \\
&= -\boldsymbol{Z}\tilde{\Phi}^*\tilde{\boldsymbol{C}}^*\tilde{\boldsymbol{C}}\tilde{\Phi}\boldsymbol{x}_0 - \boldsymbol{Z}\tilde{\Phi}^*\tilde{\boldsymbol{C}}^*\tilde{\boldsymbol{G}}\boldsymbol{w}^o
\end{aligned} \tag{A.10}$$

and

$$\begin{aligned}\boldsymbol{w}^o &= \tilde{\boldsymbol{D}}^*\tilde{\boldsymbol{D}}\boldsymbol{w}^o - \tilde{\boldsymbol{B}}^*\eta^o + \tilde{\boldsymbol{D}}^*\tilde{\boldsymbol{C}}\tilde{\boldsymbol{x}}^o \\ &= \tilde{\boldsymbol{G}}^*\tilde{\boldsymbol{G}}\boldsymbol{w}^o + \tilde{\boldsymbol{G}}^*\tilde{\boldsymbol{C}}\tilde{\Phi}\boldsymbol{x}_0 \end{aligned} \tag{A.11}$$

Solving for $\boldsymbol{w}^o$ in (A.11) and substituting into (A.10) gives η^o in terms of $\boldsymbol{x}_0$

$$\eta^o = -\boldsymbol{Z}\tilde{\Phi}^*\tilde{\boldsymbol{C}}^*(\boldsymbol{I} - \tilde{\boldsymbol{G}}\tilde{\boldsymbol{G}}^*)^{-1}\tilde{\boldsymbol{C}}\tilde{\Phi}\boldsymbol{x}_0$$

Thus $\eta_0^o = H_{20}x_0$ where

$$H_{20} := -\left(\boldsymbol{Z}\tilde{\Phi}^*\tilde{\boldsymbol{C}}^*(\boldsymbol{I} - \tilde{\boldsymbol{G}}\tilde{\boldsymbol{G}}^*)^{-1}\tilde{\boldsymbol{C}}\tilde{\Phi}\right)_{0,0}$$

Similarly, from (A.11), we get that

$$w_0^o = \left(\left(\boldsymbol{I} - \tilde{\boldsymbol{G}}^*\tilde{\boldsymbol{G}}\right)^{-1}\tilde{\boldsymbol{G}}^*\tilde{\boldsymbol{C}}\tilde{\Phi}\right)_{0,0} x_0 =: H_{10}x_0$$

We now proceed to show that we can get similar expressions for $\tilde{x}_k^o$, η_k^o and w_k^o. In order to do this, we start the system $\tilde{\boldsymbol{G}}$ at time k with "initial" condition x_k, and consider the problem of finding a worst-case input $\boldsymbol{w}^o(k) \in \boldsymbol{P}_{k-1}^\perp \ell_{2+}^{m_1}$ and optimal control input $\tilde{\boldsymbol{u}}^o(k) \in \boldsymbol{P}_{k-1}^\perp \ell_{2+}^{m_2}$ which solves

$$\sup_{\boldsymbol{w}\in\boldsymbol{P}_{k-1}^\perp\ell_{2+}^{m_1}} \inf_{\boldsymbol{u}} \left\{ \|\boldsymbol{P}_{k-1}^\perp \boldsymbol{z}\|_2^2 - \|\boldsymbol{P}_{k-1}^\perp \boldsymbol{w}\|_2^2 \right\} \tag{A.12}$$

such that $\boldsymbol{x} \in \boldsymbol{P}_{k-1}^\perp \ell_{2+}^n$. From the principle of optimality, it is clear that the worst-case input $\boldsymbol{P}_{k-1}^\perp \boldsymbol{w}^o(k) = \boldsymbol{P}_{k-1}^\perp \boldsymbol{w}^o$ and the optimal control $\boldsymbol{P}_{k-1}^\perp \tilde{\boldsymbol{u}}^o(k) = \boldsymbol{P}_{k-1}^\perp \tilde{\boldsymbol{u}}^o$ As in the $k = 0$ case,

$$\boldsymbol{P}_{k-1}^\perp \tilde{\boldsymbol{x}} = \boldsymbol{P}_{k-1}^\perp \tilde{\Phi}\boldsymbol{x}_k + \boldsymbol{P}_{k-1}^\perp \tilde{\Phi}\boldsymbol{Z}^*\tilde{\boldsymbol{B}}\boldsymbol{P}_{k-1}^\perp \boldsymbol{w}$$

This leads to the equation

$$\boldsymbol{P}_{k-1}^\perp \eta^o = -\boldsymbol{P}_{k-1}^\perp \boldsymbol{Z}\tilde{\Phi}^*\tilde{\boldsymbol{C}}^*\left(\boldsymbol{I} - \tilde{\boldsymbol{G}}\tilde{\boldsymbol{G}}^*\right)^{-1}\tilde{\boldsymbol{C}}\tilde{\Phi}\boldsymbol{x}_k$$

Now, we set

$$H_{2k} := -\left(\boldsymbol{Z}\tilde{\Phi}^*\tilde{\boldsymbol{C}}^*\left(\boldsymbol{I} - \tilde{\boldsymbol{G}}\tilde{\boldsymbol{G}}^*\right)^{-1}\tilde{\boldsymbol{C}}\tilde{\Phi}\right)_{k,k}$$

and $\boldsymbol{H}_2 = \text{diag}\{ H_{2k} \}$ giving us the desired operator. Since $\tilde{\boldsymbol{G}}$ is stable, it is immediate that $\boldsymbol{H}_2$ is bounded. A similar expression for $\boldsymbol{H}_1$ can be obtained by setting

$$H_{1k} := \left(\left(\boldsymbol{I} - \tilde{\boldsymbol{G}}^* \tilde{\boldsymbol{G}} \right)^{-1} \tilde{\boldsymbol{G}}^* \tilde{\boldsymbol{C}} \tilde{\boldsymbol{\Phi}} \right)_{k,k}$$

and $\boldsymbol{H}_1 := \text{diag}\{H_{1k}\}$.

Claim 8 *There exists a memoryless operator $\boldsymbol{X}$ satisfying the equation $\boldsymbol{Z}^* \eta^o = -\boldsymbol{X} \tilde{\boldsymbol{x}}^o + \Delta_0 \boldsymbol{f}^o$. Moreover, for this $\boldsymbol{X}$, we find $J(x_0) = x_0^T X_0 x_0$*

Proof: We have

$$\begin{aligned} \boldsymbol{Z}^* \eta^o &= \tilde{\boldsymbol{A}}^* \eta^o - \tilde{\boldsymbol{C}}^* \tilde{\boldsymbol{C}} \tilde{\boldsymbol{x}}^o - \tilde{\boldsymbol{C}}^* \tilde{\boldsymbol{D}} \boldsymbol{w}^o + \Delta_0 \boldsymbol{f}^o \\ &= (\tilde{\boldsymbol{A}}^* \boldsymbol{H}_2 - \tilde{\boldsymbol{C}}^* \tilde{\boldsymbol{C}} - \tilde{\boldsymbol{C}}^* \tilde{\boldsymbol{D}} \boldsymbol{H}_1) \tilde{\boldsymbol{x}}^o + \Delta_0 \boldsymbol{f}^o \end{aligned}$$

Hence we can define $\boldsymbol{X} = -(\tilde{\boldsymbol{A}}^* \boldsymbol{H}_2 - \tilde{\boldsymbol{C}}^* \tilde{\boldsymbol{C}} - \tilde{\boldsymbol{C}}^* \tilde{\boldsymbol{D}} \boldsymbol{H}_1)$. Furthermore, using this $\boldsymbol{X}$ we can derive

$$\begin{aligned} &\|\tilde{\boldsymbol{z}}^o\|_2^2 - \|\boldsymbol{w}^o\|_2^2 \\ &= \langle \tilde{\boldsymbol{C}} \tilde{\boldsymbol{x}}^o + \tilde{\boldsymbol{D}} \boldsymbol{w}^o, \tilde{\boldsymbol{C}} \tilde{\boldsymbol{x}}^o + \tilde{\boldsymbol{D}} \boldsymbol{w}^o \rangle - \langle \boldsymbol{w}^o, \boldsymbol{w}^o \rangle - \langle \boldsymbol{X} \tilde{\boldsymbol{x}}^o, \tilde{\boldsymbol{x}}^o \rangle \\ &\qquad + \langle \boldsymbol{X} \tilde{\boldsymbol{x}}^o, \tilde{\boldsymbol{x}}^o \rangle \\ &= \langle \tilde{\boldsymbol{C}} \tilde{\boldsymbol{x}}^o + \tilde{\boldsymbol{D}} \boldsymbol{w}^o, \tilde{\boldsymbol{C}} \tilde{\boldsymbol{x}}^o + \tilde{\boldsymbol{D}} \boldsymbol{w}^o \rangle - \langle \boldsymbol{w}^o, \boldsymbol{w}^o \rangle + x_0^T X_0 x_0 \\ &\qquad + \langle \boldsymbol{Z}^* \eta^o - \Delta_0 \boldsymbol{f}^o, \tilde{\boldsymbol{x}}^o \rangle + \langle \boldsymbol{Z} \boldsymbol{X} \tilde{\boldsymbol{x}}^o, \boldsymbol{Z} \tilde{\boldsymbol{x}}^o \rangle \\ &= \langle \tilde{\boldsymbol{C}} \tilde{\boldsymbol{x}}^o + \tilde{\boldsymbol{D}} \boldsymbol{w}^o, \tilde{\boldsymbol{C}} \tilde{\boldsymbol{x}}^o + \tilde{\boldsymbol{D}} \boldsymbol{w}^o \rangle - \langle \boldsymbol{w}^o, \boldsymbol{w}^o \rangle \\ &\qquad + \left\langle \tilde{\boldsymbol{A}}^* \eta^o - \tilde{\boldsymbol{C}}^* \tilde{\boldsymbol{C}} \tilde{\boldsymbol{x}}^o - \tilde{\boldsymbol{C}}^* \tilde{\boldsymbol{D}} \boldsymbol{w}^o, \tilde{\boldsymbol{x}}^o \right\rangle \\ &\qquad - \langle \eta^o, \tilde{\boldsymbol{A}} \tilde{\boldsymbol{x}}^o + \tilde{\boldsymbol{B}} \boldsymbol{w}^o \rangle + x_0^T X_0 x_0 \\ &= \langle \tilde{\boldsymbol{D}}^* \tilde{\boldsymbol{C}} \tilde{\boldsymbol{x}}^o + \tilde{\boldsymbol{D}}^* \tilde{\boldsymbol{D}} \boldsymbol{w}^o - \tilde{\boldsymbol{B}}^* \eta^o, \boldsymbol{w}^o \rangle - \langle \boldsymbol{w}^o, \boldsymbol{w}^o \rangle + x_0^T X_0 x_0 \\ &= x_0^T X_0 x_0 \end{aligned}$$

Claim 9 *The operator*

$$\boldsymbol{S}_2 := \boldsymbol{D}_{12}^* \boldsymbol{D}_{12} + \boldsymbol{B}_2^* \boldsymbol{Z} \boldsymbol{X} \boldsymbol{Z}^* \boldsymbol{B}_2$$

satisfies $\boldsymbol{S}_2 + \boldsymbol{S}_2^* > 0$. Moreover, the operator

$$\begin{aligned} \boldsymbol{S}_1 := \boldsymbol{I} &- \boldsymbol{D}_{11}^* \boldsymbol{D}_{11} - \boldsymbol{B}_1^* \boldsymbol{Z} \boldsymbol{X} \boldsymbol{Z}^* \boldsymbol{B}_1 \\ &+ \frac{1}{2} \boldsymbol{B}_1^* \boldsymbol{Z} (\boldsymbol{X} + \boldsymbol{X}^*) \boldsymbol{Z}^* \boldsymbol{B}_2 (\boldsymbol{S}_2 + \boldsymbol{S}_2^*)^{-1} \boldsymbol{B}_2^* \boldsymbol{Z} (\boldsymbol{X} + \boldsymbol{X}^*) \boldsymbol{Z}^* \boldsymbol{B}_1 \end{aligned}$$

satisfies $\boldsymbol{S}_1 + \boldsymbol{S}_1^* > 0$.

Proof: We do not know yet if $\boldsymbol{X}$ is symmetric, but from the definition we can see that it is a diagonal operator, hence we can write $\boldsymbol{X} = \operatorname{diag}\{X_k\}$. By Claim 4 and Claim 8 we know that $X_0 + X_0^* \geq 2L_0$. Using similar arguments and the principle of optimality applied to (A.12), we see that $X_k + X_k^* \geq 2L_k$. Hence $\boldsymbol{S}_2$ satisfies

$$\boldsymbol{S}_2 + \boldsymbol{S}_2^* \geq 2(\boldsymbol{D}_{12}^*\boldsymbol{D}_{12} + \boldsymbol{B}_2^*\boldsymbol{Z}\boldsymbol{L}\boldsymbol{Z}^*\boldsymbol{B}_2) = 2\boldsymbol{N} > \boldsymbol{0}$$

which we know from equation (A.3). Since we know that there exists a feedback such that the closed-loop norm is less than 1, there exists a $\delta > 0$ such that, if $x_0 = 0$, for all $\boldsymbol{w} \in \ell_{2+}^{m_1}$

$$\inf_{u}\left\{ \|\boldsymbol{z}\|_2^2 - \|\boldsymbol{w}\|_2^2 \right\} \leq -\delta^2 \|\boldsymbol{w}\|_2^2 \tag{A.13}$$

Take any non-negative integer k and $\hat{w} \in \mathbb{R}^{m_1}$. We take $\boldsymbol{P}_{k-1}\boldsymbol{w} = \boldsymbol{0}$ and $w_k = \hat{w}$. Since $x_0 = 0$ and $\boldsymbol{u}$ is a causal feedback, the optimal control must satisfies $\boldsymbol{P}_{k-1}\boldsymbol{u} = \boldsymbol{0}$ and $\boldsymbol{P}_{k-1}\boldsymbol{x} = \boldsymbol{0}$. The optimal u_k only depends on $\boldsymbol{P}_k\boldsymbol{w}$, hence we get

$$\begin{aligned}
&\sup_{\boldsymbol{w}\in \boldsymbol{P}_k^{\perp}\ell_{2+}^{m_1}} \inf_{\boldsymbol{u}} \left\{ \|\boldsymbol{z}\|_2^2 - \|\boldsymbol{w}\|_2^2 \,\middle|\, \boldsymbol{u} \in \boldsymbol{P}_{k-1}^{\perp}\ell_{2+}^{m_2} \text{ such that } \boldsymbol{x} \in \boldsymbol{P}_{k-1}^{\perp}\ell_{2+}^{n} \right\} \\
&= \inf_{u_k} \sup_{\boldsymbol{w}\in \boldsymbol{P}_k^{\perp}\ell_{2+}^{m_1}} \inf_{\boldsymbol{u}} \left\{ \|\boldsymbol{z}\|_2^2 - \|\boldsymbol{w}\|_2^2 \,\middle|\, \boldsymbol{u} \in \boldsymbol{P}_{k}^{\perp}\ell_{2+}^{m_2} \text{ such that } \boldsymbol{x} \in \boldsymbol{P}_{k}^{\perp}\ell_{2+}^{n} \right\} \\
&= \inf_{u_k} \left\{ |z_k|^2 - |\hat{w}|^2 + x_{k+1}^T X_{k+1} x_{k+1} \right\} \\
&= \inf_{u_k} \begin{bmatrix} \hat{w} \\ u_k \end{bmatrix}^T \Gamma \begin{bmatrix} \hat{w} \\ u_k \end{bmatrix}
\end{aligned} \tag{A.14}$$

where

$$\Gamma := \begin{bmatrix} (\boldsymbol{D}_{11}^*\boldsymbol{D}_{11} + \boldsymbol{B}_1^*\boldsymbol{Z}\boldsymbol{X}\boldsymbol{Z}^*\boldsymbol{B}_1 - \boldsymbol{I})_{k,k} & (\boldsymbol{B}_1^*\boldsymbol{Z}\boldsymbol{X}\boldsymbol{Z}^*\boldsymbol{B}_2)_{k,k} \\ (\boldsymbol{B}_2^*\boldsymbol{Z}\boldsymbol{X}\boldsymbol{Z}^*\boldsymbol{B}_1)_{k,k} & (\boldsymbol{D}_{12}^*\boldsymbol{D}_{12} + \boldsymbol{B}_2^*\boldsymbol{Z}\boldsymbol{X}\boldsymbol{Z}^*\boldsymbol{B}_2)_{k,k} \end{bmatrix}$$

Continuing, we have that

$$\begin{aligned}
\text{(A.14)} &= \frac{1}{2}\inf_{\bar{u}_k} \begin{bmatrix} \hat{w} \\ \bar{u}_k \end{bmatrix}^T \begin{bmatrix} -(\boldsymbol{S}_1 + \boldsymbol{S}_1^*)_{k,k} & \boldsymbol{0} \\ \boldsymbol{0} & (\boldsymbol{S}_2 + \boldsymbol{S}_2^*)_{k,k} \end{bmatrix} \begin{bmatrix} \hat{w} \\ \bar{u}_k \end{bmatrix} \\
&= -\frac{1}{2}\hat{w}^T(\boldsymbol{S}_1 + \boldsymbol{S}_1^*)_{k,k}\hat{w} \\
&\leq -\delta^2 \|\boldsymbol{w}\|_2^2 \\
&\leq -\delta^2 \hat{w}^T\hat{w}
\end{aligned}$$

where $\bar{u}_k := u_k + \left((S_2 + S_2^*)^{-1}(B_2^* Z(X + X^*)Z^* B_1)\right)_{k,k} \hat{w}$. The first inequality is a result of (A.13). Since $\hat{w} \in \mathbb{R}^{m_1}$ is arbitrary, this implies that $-(S_1 + S_1^*)_{k,k} \le -2\delta^2 I$ for every k, hence $S_1 + S_1^* > 0$.

Claim 10 *The operator X is such that A_{cl} is UES, where*

$$A_{cl} := A - B(R + B^* Z X Z^* B)^{-1}(B^* Z X Z^* A + D_{1\bullet}^* C_1)$$

Proof: Using Claim 8 we can write

$$\begin{aligned}\eta^o &= Z Z^* \eta^o \\ &= -Z X \tilde{x}^o + Z \Delta_0 f^o \\ &= -Z X Z^* Z \tilde{x}^o - Z X \Delta_0 \tilde{x}^o \\ &= -Z X Z^*(A \tilde{x}^o + B_1 w^o + B_2 \tilde{u}^o)\end{aligned}$$

since $Z\Delta_0 = ZX\Delta_0 = 0$. By substituting this in the equations

$$\begin{aligned}(I - D_{11}^* D_{11}) w^o &= -B_1^* \eta^o + D_{11}^* C_1 \tilde{x}^o \\ D_{12}^* D_{12} \tilde{u}^o &= B_2^* \eta^o\end{aligned}$$

we can rewrite this as

$$(R + B^* Z X Z^* B) \begin{bmatrix} w^o \\ \tilde{u}^o \end{bmatrix} = -(B^* Z X Z^* A + D_{1\bullet}^* C_1) \tilde{x}^o \tag{A.15}$$

Since $S_1 + S_1^*$ and $S_2 + S_2^*$ both have a bounded inverse, it can be shown that $R + B^* Z X Z^* B$ also has a bounded inverse. Hence the closed-loop system satisfies $Z \tilde{x}^o = A_{cl} \tilde{x}^o$, where A_{cl} can be written in the feedback form $A + BF$ for

$$F = -(R + B^* Z X Z^* B)^{-1}(B^* Z X Z^* A + D_{1\bullet}^* C_1)$$

Since

$$\begin{bmatrix} w^o \\ \tilde{u}^o \end{bmatrix} = F \Phi_{cl} x_0 \in \ell_{2+}^{m_1} \oplus \ell_{2+}^{m_2}, \quad \text{where } \Phi_{cl} := (I - Z^* A_{cl})^{-1}$$

for arbitrary $x_0 \in P_0 \ell_{2+}^n$, it follows that $F\Phi_{cl}$ is in $\mathcal{B}(\ell_{2+}^n, \ell_{2+}^{m_1} \oplus \ell_{2+}^{m_2})$. Since (A_{cl}, I) and (F, A_{cl}) are uniformly exponentially stabilizable and detectable, we may use Lemma 2.10 completing the proof that A_{cl} is UES.

Before showing that X is positive semi-definite, we will show that it satisfies the Riccati operator equation.

Claim 11 *The operator $\boldsymbol{X}$ satisfies the operator Riccati equation (6.3).*

Proof: Using Claim 8 and equations (A.5) and (A.15) we find

$$\begin{aligned}
\boldsymbol{X}\tilde{\boldsymbol{x}}^o &= -\boldsymbol{Z}^*\eta^o + \Delta_0 \boldsymbol{f}^o \\
&= -\boldsymbol{A}^*\boldsymbol{Z}\boldsymbol{Z}^*\eta^o + \boldsymbol{C}_1^*\boldsymbol{C}_1\tilde{\boldsymbol{x}}^o + \boldsymbol{C}_1^*\boldsymbol{D}_{11}\boldsymbol{w}^o \\
&= \boldsymbol{A}^*\boldsymbol{Z}\boldsymbol{X}\tilde{\boldsymbol{x}}^o + \boldsymbol{C}_1^*\boldsymbol{C}_1\tilde{\boldsymbol{x}}^o + \boldsymbol{C}_1^*\boldsymbol{D}_{1\bullet}\begin{bmatrix} \boldsymbol{w}^o \\ \tilde{\boldsymbol{u}}^o \end{bmatrix} - \boldsymbol{A}^*\boldsymbol{Z}\Delta_0 \boldsymbol{f}^o \\
&= \boldsymbol{A}^*\boldsymbol{Z}\boldsymbol{X}\boldsymbol{Z}^*\boldsymbol{Z}\tilde{\boldsymbol{x}}^o + \boldsymbol{C}_1^*\boldsymbol{C}_1\tilde{\boldsymbol{x}}^o + \boldsymbol{C}_1^*\boldsymbol{D}_{1\bullet}\boldsymbol{F}\tilde{\boldsymbol{x}}^o + \boldsymbol{A}^*\boldsymbol{Z}\boldsymbol{X}\Delta_0\tilde{\boldsymbol{x}}^o \\
&= \boldsymbol{A}^*\boldsymbol{Z}\boldsymbol{X}\boldsymbol{Z}^*(\boldsymbol{A} + \boldsymbol{B}\boldsymbol{F})\tilde{\boldsymbol{x}}^o + \boldsymbol{C}_1^*\boldsymbol{C}_1\tilde{\boldsymbol{x}}^o + \boldsymbol{C}_1^*\boldsymbol{D}_{1\bullet}\boldsymbol{F}\tilde{\boldsymbol{x}}^o \\
&= \big(\boldsymbol{A}^*\boldsymbol{Z}\boldsymbol{X}\boldsymbol{Z}^*\boldsymbol{A} + \boldsymbol{C}_1^*\boldsymbol{C}_1 + (\boldsymbol{A}^*\boldsymbol{Z}\boldsymbol{X}\boldsymbol{Z}^*\boldsymbol{B} + \boldsymbol{C}_1^*\boldsymbol{D}_{1\bullet})\boldsymbol{F}\big)\tilde{\boldsymbol{x}}^o
\end{aligned}$$

which is true for all x_0. Since the operators are all memoryless, it follows that

$$\Big(\boldsymbol{X} - \boldsymbol{A}^*\boldsymbol{Z}\boldsymbol{X}\boldsymbol{Z}^*\boldsymbol{A} - \boldsymbol{C}_1^*\boldsymbol{C}_1 - (\boldsymbol{A}^*\boldsymbol{Z}\boldsymbol{X}\boldsymbol{Z}^*\boldsymbol{B} + \boldsymbol{C}_1^*\boldsymbol{D}_{1\bullet})\boldsymbol{F}\Big)_{0,0} = 0$$

By starting the system at time k with "initial" condition x_k, it follows from the principle of optimality that for every k

$$\Big(\boldsymbol{X} - \boldsymbol{A}^*\boldsymbol{Z}\boldsymbol{X}\boldsymbol{Z}^*\boldsymbol{A} - \boldsymbol{C}_1^*\boldsymbol{C}_1 - (\boldsymbol{A}^*\boldsymbol{Z}\boldsymbol{X}\boldsymbol{Z}^*\boldsymbol{B} + \boldsymbol{C}_1^*\boldsymbol{D}_{1\bullet})\boldsymbol{F}\Big)_{k,k} = 0$$

which gives the operator Riccati equation.

Claim 12 *The operator $\boldsymbol{X}$ satisfies $\boldsymbol{X} \geq \boldsymbol{0}$.*

Proof: By subtracting the adjoint of the Riccati operator equation (6.3) from the equation itself, it is straightforward that

$$\boldsymbol{X} - \boldsymbol{X}^* = \boldsymbol{A}_{cl}^*\boldsymbol{Z}(\boldsymbol{X} - \boldsymbol{X}^*)\boldsymbol{Z}^*\boldsymbol{A}_{cl}$$

Since $\boldsymbol{A}_{cl}$ is UES, it follows from Lemma 2.8 that $\boldsymbol{X} = \boldsymbol{X}^*$. In Claim 9 we have seen that $\boldsymbol{X} + \boldsymbol{X}^* \geq 2\boldsymbol{L}$; thus we have $\boldsymbol{X} \geq \boldsymbol{L} \geq \boldsymbol{0}$.

Claim 13 *There exist memoryless operators $\boldsymbol{T}_{11}$, $\boldsymbol{T}_{21}$ and $\boldsymbol{T}_{22}$ with $\boldsymbol{T}_{11} > \boldsymbol{0}$ and $\boldsymbol{T}_{22} > \boldsymbol{0}$ such that*

$$\begin{aligned}
&\begin{bmatrix} \boldsymbol{D}_{11}^*\boldsymbol{D}_{11} + \boldsymbol{B}_1^*\boldsymbol{Z}\boldsymbol{X}\boldsymbol{Z}^*\boldsymbol{B}_1 - \boldsymbol{I} & \boldsymbol{B}_1^*\boldsymbol{Z}\boldsymbol{X}\boldsymbol{Z}^*\boldsymbol{B}_2 \\ \boldsymbol{B}_2^*\boldsymbol{Z}\boldsymbol{X}\boldsymbol{Z}^*\boldsymbol{B}_1 & \boldsymbol{D}_{12}^*\boldsymbol{D}_{12} + \boldsymbol{B}_2^*\boldsymbol{Z}\boldsymbol{X}\boldsymbol{Z}^*\boldsymbol{B}_2 \end{bmatrix} \\
&\qquad = \begin{bmatrix} \boldsymbol{T}_{21}^*\boldsymbol{T}_{21} - \boldsymbol{T}_{11}^*\boldsymbol{T}_{11} & \boldsymbol{T}_{21}^*\boldsymbol{T}_{22} \\ \boldsymbol{T}_{22}^*\boldsymbol{T}_{21} & \boldsymbol{T}_{22}^*\boldsymbol{T}_{22} \end{bmatrix}
\end{aligned}$$

Since $\boldsymbol{X} = \boldsymbol{X}^*$ we get

$$\begin{aligned} \boldsymbol{S}_2 &= \boldsymbol{D}_{12}^*\boldsymbol{D}_{12} + \boldsymbol{B}_2^*\boldsymbol{Z}\boldsymbol{X}\boldsymbol{Z}^*\boldsymbol{B}_2 > \boldsymbol{0} \\ \boldsymbol{S}_1 &= \boldsymbol{I} - \boldsymbol{D}_{11}^*\boldsymbol{D}_{11} - \boldsymbol{B}_1^*\boldsymbol{Z}\boldsymbol{X}\boldsymbol{Z}^*\boldsymbol{B}_1 \\ &\qquad + \boldsymbol{B}_1^*\boldsymbol{Z}\boldsymbol{X}\boldsymbol{Z}^*\boldsymbol{B}_2\boldsymbol{S}_2^{-1}\boldsymbol{B}_2^*\boldsymbol{Z}\boldsymbol{X}\boldsymbol{Z}^*\boldsymbol{B}_1 > \boldsymbol{0} \end{aligned}$$

Both operators are memoryless and positive, hence they admit memoryless positive square roots. From this it is obvious that we can define

$$\begin{aligned} \boldsymbol{T}_{22} &:= (\boldsymbol{D}_{12}^*\boldsymbol{D}_{12} + \boldsymbol{B}_2^*\boldsymbol{Z}\boldsymbol{X}\boldsymbol{Z}^*\boldsymbol{B}_2)^{1/2} \\ \boldsymbol{T}_{21} &:= \boldsymbol{T}_{22}^{-*}(\boldsymbol{B}_2^*\boldsymbol{Z}\boldsymbol{X}\boldsymbol{Z}^*\boldsymbol{B}_1) \\ \boldsymbol{T}_{11} &:= \Big(\boldsymbol{I} - \boldsymbol{D}_{11}^*\boldsymbol{D}_{11} - \boldsymbol{B}_1^*\boldsymbol{Z}\boldsymbol{X}\boldsymbol{Z}^*\boldsymbol{B}_1 \\ &\qquad + (\boldsymbol{B}_1^*\boldsymbol{Z}\boldsymbol{X}\boldsymbol{Z}^*\boldsymbol{B}_2)\boldsymbol{T}_{22}^{-1}\boldsymbol{T}_{22}^{-*}(\boldsymbol{B}_2^*\boldsymbol{Z}\boldsymbol{X}\boldsymbol{Z}^*\boldsymbol{B}_1)\Big)^{1/2} \end{aligned}$$

which completes the proof. ■

Remark A.1 If the assumption (A.1) is not satisfied, we first apply a feedback

$$\boldsymbol{u} = -(\boldsymbol{D}_{12}^*\boldsymbol{D}_{12})^{-1}\boldsymbol{D}_{12}^*\boldsymbol{C}_1\boldsymbol{x} - (\boldsymbol{D}_{12}^*\boldsymbol{D}_{12})^{-1}\boldsymbol{D}_{12}^*\boldsymbol{D}_{11}\boldsymbol{w} + \boldsymbol{v}$$

which is well-defined since $\boldsymbol{D}_{12}^*\boldsymbol{D}_{12} > \boldsymbol{0}$. For the resulting plant assumption (A.1) is satisfied, as well as the assumption that the pair $(\boldsymbol{C}_1, \boldsymbol{A})$ is uniformly detectable. Hence we find solutions say $\bar{\boldsymbol{X}}$, $\bar{\boldsymbol{T}}$, and the corresponding $\bar{\boldsymbol{A}}_{cl}$. By rewriting the equations it can be checked that the original plant has solutions $\boldsymbol{X} = \bar{\boldsymbol{X}}$ and

$$\boldsymbol{T} = \bar{\boldsymbol{T}}\begin{bmatrix} \boldsymbol{I} & 0 \\ (\boldsymbol{D}_{12}^*\boldsymbol{D}_{12})^{-1}\boldsymbol{D}_{12}^*\boldsymbol{D}_{11} & \boldsymbol{I} \end{bmatrix}$$

and that $\boldsymbol{A}_{cl} = \bar{\boldsymbol{A}}_{cl}$.

Proof of Theorem 7.21 B

To evaluate $R(G, -\gamma^{-2})$, we follow the lines of [39], who considered the analogous result for time-invariant systems defined on an infinite horizon. Let H be the covariance matrix of the process z, i.e.

$$\mathcal{E} z_t z_s^T = H_{t,s} = \int_{-T}^{\min\{t,s\}} g_{t,\tau} g_{s,\tau}^T \, d\tau$$

It follows from Mercer's theorem [17] that the integral equation

$$\int_{-T}^{T} H_{t,s} \phi_{i_s} \, ds = \lambda_i \phi_{i_t} \tag{B.1}$$

has solutions ϕ_i as a function of t. In fact there are at most a countable number of distinct solutions, which are the eigenfunctions of the integral operator with kernel $H_{t,s}$. We will normalize the eigenfunctions, i.e.

$$\int_{-T}^{T} \phi_{i_t}^T \phi_{i_t} \, dt = 1$$

Assuming that $H_{t,s}$ is strictly positive definite, these eigenfunctions can be chosen to form a complete orthonormal set (in the $\mathcal{L}_2$ sense) over the interval $[-T, T]$.

When $H_{t,s}$ is not strictly positive definite, we may augment the set with enough additional orthonormal functions to obtain a complete set. By the Karhunen-Loève theorem [17], the signal z can be expanded as

$$z_t = \sum_{i=1}^{\infty} x_i \phi_{i_t} \tag{B.2}$$

where

$$x_i = \int_{-T}^{T} z_t^T \phi_{i_t}\, dt$$

Here the convergence of the infinite sum is to be understood as mean square convergence. Furthermore, the $\{x_i\}_{i=1,2,\cdots}$ are zero mean independent Gaussian random variables with

$$\mathcal{E} x_i^2 = \lambda_i \quad i = 1, 2, \cdots$$

It follows that the processes in (B.2) have the same probability density function, and hence

$$\int_{-T}^{T} z_t^T z_t\, dt = \int_{-T}^{T} \sum_{i=1}^{\infty} x_i \phi_{i_t}^T \sum_{j=1}^{\infty} x_j \phi_{j_t}\, dt = \sum_{i=1}^{\infty} x_i^2$$

(by orthogonality of ϕ_{i_t} on $[-T, T]$) have the same distribution. Therefore, we can evaluate as in [39]

$$\begin{aligned}
R(G, -\gamma^{-2}) &= 2\gamma^2 \ln \mathcal{E} \exp\Big(\frac{1}{2\gamma^2} \int_{-T}^{T} |z_t|^2\, dt\Big) \\
&= 2\gamma^2 \ln \mathcal{E} \exp\Big(\frac{1}{2\gamma^2} \sum_{i=1}^{\infty} x_i^2\Big) \\
&= 2\gamma^2 \ln \prod_{i=1}^{\infty} \mathcal{E} \exp\Big(\frac{1}{2\gamma^2} x_i^2\Big) \\
&= 2\gamma^2 \sum_{i=1}^{\infty} \ln \frac{\lambda_i^{-1/2}}{\sqrt{2\pi}} \int_{-\infty}^{\infty} \exp\left(-\frac{x_i^2}{2}(\lambda_i^{-1} - \gamma^{-2})\right) \\
&= 2\gamma^2 \sum_{i=1}^{\infty} \ln(\lambda_i^{-1} - \gamma^{-2})^{-1/2} \lambda_i^{-1/2} \\
&= -\gamma^2 \sum_{i=1}^{\infty} \ln(1 - \lambda_i \gamma^{-2})
\end{aligned} \tag{B.3}$$

At this point we must depart the approach used in [39], where the eigenvalues are estimated in terms of the transfer function. In order to estimate the eigenvalues λ_i, we will use the results that were obtained in the discrete-time case. We sample and hold the input, as well as sampling the output. To ease the notation we denote $r(k) = -T + k\frac{2T}{N}$. For a signal w acting on an interval $[-T, T]$, we define

$$\bar{w}_t := w_{r(k)} \quad t \in [r(k), r(k+1))$$

for $k = 0, 2, \cdots, N-1$. For the input $\bar{w}$, the sampled output equals

$$z_{r(k)} = \int_{-T}^{r(k)} g_{r(k),\tau} w_\tau \, d\tau = \sum_{j=0}^{k-1} G_{k,j} w_{r(j)}$$

where

$$G_{k,j} := \int_{r(j)}^{r(j+1)} g_{(k),\tau} \, d\tau$$

Now, we define the finite horizon discrete-time mapping $\tilde{G}$ as $G_{k,j}$ for $0 \leq j < k \leq N-1$. We can rewrite (B.1) as

$$\lim_{N\to\infty} \frac{2T}{N} \sum_{j=0}^{N-1} H_{t,r(j)} \phi_{i_{r(j)}} = \lambda_i \phi_{i_t}$$

At $t = r(k)$ for $k = \{0, \cdots, N-1\}$ this gives

$$\begin{aligned}
\lambda_i \phi_{i_{r(k)}} &= \lim_{N\to\infty} \frac{2T}{N} \sum_{j=0}^{N-1} H_{r(k),r(j)} \phi_{i_{r(j)}} \\
&= \lim_{N\to\infty} \frac{2T}{N} \sum_{j=0}^{N-1} \int_{-T}^{\min\{r(k),r(j)\}} g_{r(k),\tau} g_{r(j),\tau}^T \, d\tau \, \phi_{i_{r(j)}} \\
&= \lim_{N\to\infty} \frac{2T}{N} \sum_{j=0}^{N-1} \sum_{p=0}^{\min\{k,j\}-1} \int_{r(p)}^{r(p+1)} g_{r(k),\tau} g_{r(j),\tau}^T \, d\tau \, \phi_{i_{r(j)}} \\
&= \lim_{N\to\infty} \sum_{j=0}^{N-1} \left(\sum_{p=0}^{\min\{k,j\}-1} G_{k,p} G_{j,p}^T \, d\tau \right. \\
&\quad \left. - \sum_{p=0}^{\min\{k,j\}-1} \int_{r(p)}^{r(p+1)} g_{r(k),\tau} \left(G_{j,p}^T \frac{2T}{N} g_{r(j),\tau}^T \right) d\tau \right) \phi_{i_{r(j)}}
\end{aligned}$$

Using continuity arguments it can be shown that the second term can be made arbitrarily small. Thus

$$\lim_{N\to\infty} \sum_{j=0}^{N-1} (\tilde{G}\tilde{G}^T)_{k,j}\phi_{i_{r(j)}} = \lambda_i \phi_{i_{r(k)}}$$

Now, by defining $\phi_i = (\phi^T_{i_{r(0)}} \quad \cdots \quad \phi^T_{i_{r(N-1)}})^T$, we see that

$$\lim_{N\to\infty} \tilde{G}\tilde{G}^T \phi_i = \lambda_i \phi_i$$

We are interested in the eigenvalues of $\tilde{G}\tilde{G}^T$, which are the same (except for zero eigenvalues) as those of $\tilde{G}^T\tilde{G}$. Suppose that $\tilde{L}$ is a spectral factor, i.e.

$$I - \gamma^{-2}\tilde{G}^T\tilde{G} = \tilde{L}^T\tilde{L}$$

Then we have

$$\lim_{N\to\infty} \tilde{L}^T\tilde{L}\phi_i = \lim_{N\to\infty} (1 - \gamma^{-2}\tilde{G}^T\tilde{G})\phi_i = (1 - \gamma^{-2}\lambda_i)\phi_i$$

Denote the eigenvalues of $\tilde{L}^T\tilde{L}$ by $\bar{\lambda}_i$ for $i = 1, \cdots, Nm$. We can evaluate the finite LEQG cost using (B.3) as

$$\begin{aligned}
R(G, -\gamma^{-2}) &= -\gamma^2 \sum_{i=1}^{\infty} \ln(1 - \lambda_i\gamma^{-2}) \\
&= -\gamma^2 \lim_{N\to\infty} \sum_{i=1}^{Nm} \ln \bar{\lambda}_i \\
&= -\gamma^2 \lim_{N\to\infty} \ln \prod_{i=1}^{Nm} \bar{\lambda}_i \\
&= -\gamma^2 \lim_{N\to\infty} \ln \det(\tilde{L}^T\tilde{L})
\end{aligned}$$

Since $\tilde{L}$ is a lower block triangular matrix, this can be written as

$$\begin{aligned}
R(G, -\gamma^{-2}) &= -2\gamma^2 \lim_{N\to\infty} \ln \det(\tilde{L}) \\
&= -2\gamma^2 \lim_{N\to\infty} \ln \prod_{i=0}^{N-1} \det(L_{i,i}) \\
&= -2\gamma^2 \lim_{N\to\infty} \sum_{i=0}^{N-1} \ln \det(L_{i,i}) \\
&= -\gamma^2 \lim_{N\to\infty} \sum_{i=0}^{N-1} \ln \det(L^T_{i,i}L_{i,i})
\end{aligned}$$

At this point we would like to relate the memoryless part of $\tilde{L}$ to the memoryless part of the continuous time spectral factor L in (7.27). This can be done as in Theorem 7.14, resulting in

$$\lim_{N\to\infty} L_{k,k}^T L_{k,k} = I + \lim_{N\to\infty} \frac{2T}{N}\left(\frac{l_{r(k),r(k)}}{2} + \frac{l_{r(k),r(k)}^T}{2}\right)$$

Using this, the finite horizon LEQG cost equals

$$\begin{aligned}
R(G,-\gamma^{-2}) &= -\gamma^2 \lim_{N\to\infty} \sum_{i=0}^{N-1} \ln\det(L_{i,i}^T L_{i,i}) \\
&= -\gamma^2 \lim_{N\to\infty} \sum_{i=0}^{N-1} \ln\det\left(I + \frac{2T}{N}\left(\frac{l_{r(i),r(i)}}{2} + \frac{l_{r(i),r(i)}^T}{2}\right)\right) \\
&= -\gamma^2 \lim_{N\to\infty} \frac{2T}{N} \sum_{i=0}^{N-1} \frac{N}{2T} \\
&\quad \times \ln\det\left(I + \frac{2T}{N}\left(\frac{l_{r(i),r(i)}}{2} + \frac{l_{r(i),r(i)}^T}{2}\right)\right)
\end{aligned}$$

For fixed N, we can use Lemma 7.1 to write

$$\begin{aligned}
\frac{N}{2T} \ln\det\left(I + \frac{2T}{N}\left(\frac{l_{r(i),r(i)}}{2} + \frac{l_{r(i),r(i)}^T}{2}\right)\right) \\
= \operatorname{tr}\left[\left(\frac{l_{r(i),r(i)}}{2} + \frac{l_{r(i),r(i)}^T}{2}\right)\right] + \mathcal{O}\left(\frac{2T}{N}\right)
\end{aligned}$$

Hence

$$\begin{aligned}
&R(G,-\gamma^{-2}) \\
&= -\gamma^2 \lim_{N\to\infty} \frac{2T}{N} \sum_{i=0}^{N-1} \left(\operatorname{tr}\left[\left(\frac{l_{r(i),r(i)}}{2} + \frac{l_{r(i),r(i)}^T}{2}\right)\right] + \mathcal{O}\left(\frac{2T}{N}\right)\right) \\
&= \lim_{N\to\infty} \frac{2T}{N} \sum_{i=0}^{N-1} E(G,\gamma,r(i)) + \lim_{N\to\infty} \mathcal{O}\left(\frac{2T}{N}\right) \\
&= \int_{-T}^{T} E(G,\gamma,t)\, dt
\end{aligned}$$

where we used the expression (7.28) for the finite horizon entropy. ∎

Bibliography

[1] D. Alpay, P. Dewilde, and H. Dym. Lossless inverse scattering and reproducing kernels for upper triangular operators. In I. Gohberg, editor, *Extension and Interpolation of Linear Operators and Matrix Functions*, volume 47 of *Operator Theory, Advances and Applications*, pages 61–135. Birkhäuser, Basel, 1990.

[2] B.D.O. Anderson. Internal and external stability of linear time-varying systems. *SIAM Journal on Control and Optimization*, 20(3):408–413, May 1982.

[3] B.D.O. Anderson and J.B. Moore. Detectability and stabilizability of time-varying discrete-time linear systems. *SIAM Journal on Control and Optimization*, 19(1):20–32, January 1981.

[4] B.D.O. Anderson and S. Vongpanitlerd. *Network Analysis and Synthesis: A Modern System Theory Approach.* Prentice-Hall, Englewood Cliffs, NJ, 1973.

[5] N. Aoki, S. Ushida, and H. Kimura. On γ-characteristic of H^∞ control systems. In *Proceedings of IEEE Conference on Decision and Control*, pages 2562–2567, New Orleans, LA, December 1995.

[6] D.Z. Arov and M.G. Kreĭn. Problem of search of the minimum entropy in indeterminate extension problems. *Functional Analysis and its Applications*, 15(2):123–126, October 1981.

[7] D.Z. Arov and M.G. Kreĭn. On the evaluation of entropy functionals and their minima in generalized extension problems. *Acta Scientiarum Mathematicarum*, 45:33–50, 1983. In Russian.

[8] W. Arveson. Interpolation in nest algebras. *Journal of Functional Analysis*, 20(3):208–233, November 1975.

[9] E. Baeyens and P.P. Khargonekar. Some examples in mixed $\mathcal{H}_2/\mathcal{H}_\infty$ control. In *Proceedings of American Control Conference*, pages 1608–1612, Baltimore, MD, 1994.

[10] J.A. Ball and A.J. van der Schaft. J-inner-outer factorization, J-spectral factorization, and robust control for nonlinear systems. *IEEE Transactions on Automatic Control*, 41(3):379–392, March 1996.

[11] B. Bamieh and J.B. Pearson. A general framework for linear periodic systems with the application to $\mathcal{H}_\infty$ sampled-data control. *IEEE Transactions on Automatic Control*, 37(4):418–435, April 1992.

[12] B. Bamieh, J.B. Pearson, B.A. Francis, and A. Tannenbaum. A lifting technique for linear periodic systems with applications to sampled-data control. *Systems & Control Letters*, 17(2):79–88, August 1992.

[13] T. Başar and P. Bernhard. *$\mathcal{H}_\infty$-Optimal Control and Related Minimax Design Problems.* Birkhäuser, Boston, MA, 2nd edition, 1995.

[14] B. Beauzamy. *Introduction to Operator Theory and Invariant Subspaces.* North-Holland, Amsterdam, 1988.

[15] D.S. Bernstein and W.M. Haddad. LQG control with an $\mathcal{H}_\infty$ performance bound: A Riccati equation approach. *IEEE Transactions on Automatic Control*, 34(3):293–305, March 1989.

[16] S.P. Boyd, V. Balakrishnan, and P.T. Kabamba. A bisection method for computing the $\mathcal{H}_\infty$ norm of a transfer matrix and related problems. *Mathematics of Control, Signals and Systems*, 3(2):207–220, 1989.

[17] J.A. Bucklew. *Large Deviation Techniques in Decision, Simulation, and Estimation.* Wiley-InterScience, New York, 1990.

[18] J.P. Burg. Maximum entropy spectral analysis. In D.G. Childers, editor, *Modern Spectrum Analysis*, pages 34–41. IEEE Press, New York, 1978. Reprinted from *Proceedings of the 37th Meeting of the Society of Exploration Geophysicists*, Oklahoma City, OK, October 1967.

[19] T. Chen and B.A. Francis. On the $\mathcal{L}_2$-induced norm of a sampled-data system. *Systems & Control Letters*, 15(3):211–219, September 1990.

[20] T. Chen and B.A. Francis. Input-output stability of sampled-data systems. *IEEE Transactions on Automatic Control*, 36(1):50–58, January 1991.

[21] T. Chen and B.A. Francis. *Optimal Sampled-Data Control Systems.* Springer-Verlag, London, 1995.

[22] J. Chover. On normalized entropy and the extensions of a positive-definite function. *Journal of Mathematics and Mechanics*, 10(6):927–945, 1961.

[23] T.M. Cover and J.A. Thomas. *Elements of Information Theory.* John-Wiley and Sons, New York, 1991.

[24] W.N. Dale and M.C. Smith. Stabilizability and existence of system representations for discrete-time time-varying systems. *SIAM Journal on Control and Optimization*, 31(6):1538–1557, November 1993.

[25] K.R. Davidson. *Nest Algebras*, volume 191 of *Pitman research notes in mathematics.* Longman Scientific & Technical, Harlow, UK, 1988.

[26] C. Davis, W.M. Kahan, and H.F. Weinberger. Norm-preserving dilations and their applications to optimal error bounds. *SIAM Journal on Numerical Analysis*, 19(3):445–469, June 1982.

[27] G. de Nicolao. On the time-varying Riccati difference equation of optimal filtering. *SIAM Journal on Control and Optimization*, 30(6):1251–1269, November 1992.

[28] C.A. Desoer. Slowly varying discrete system $x_{i+1} = A_i x_i$. *Electronics Letters*, 6(11):339–340, May 1970.

[29] C.A. Desoer and M. Vidyasagar. *Feedback Systems: Input-Output Properties.* Academic Press, New York, 1975.

[30] J.C. Doyle. Guaranteed margins for LQG regulators. *IEEE Transactions on Automatic Control*, 23(4):756–757, July 1978.

[31] J.C. Doyle, K. Glover, P.P. Khargonekar, and B.A. Francis. State-space solutions to standard $\mathcal{H}_2$ and $\mathcal{H}_\infty$ control problems. *IEEE Transactions on Automatic Control*, 34(8):831–847, August 1989.

[32] H. Dym. *J-Contractive Matrix Functions, Reproducing Kernel Hilbert Spaces and Interpolation*, volume 71 of *Regional conference series in mathematics.* Conference Board of the Mathematics Sciences, American Mathematics Society, Providence, RI, 1989.

[33] H. Dym and I. Gohberg. Extensions of band matrices with band inverses. *Linear Algebra and its Applications*, 36:1–24, March 1981.

[34] H. Dym and I. Gohberg. A maximum entropy principle for contractive interpolants. *Journal of Functional Analysis*, 65(1):83–125, January 1986.

[35] A. Feintuch and R. Saeks. *System Theory: A Hilbert Space Approach.* Academic Press, New York, 1982.

[36] A. Feintuch, R. Saeks, and C. Neil. A new performance measure for stochastic optimization in Hilbert space. *Mathematical Systems Theory*, 15(1):39–54, December 1981.

[37] B.A. Francis. *A Course in $\mathcal{H}_\infty$ Control Theory*, volume 88 of *Lecture Notes in Control and Information Sciences.* Springer-Verlag, Berlin, 1987.

[38] B. Friedland, J. Richman, and D.E. Williams. On the 'adiabiatic approximation' for design of control laws for linear, time-varying systems. *IEEE Transactions on Automatic Control*, 32(1):62–63, January 1987. Corrected by I. Troch, 33(3) 318–319, March 1988.

[39] K. Glover. Minimum entropy and risk-sensitive control: the continuous time case. In *Proceedings of IEEE Conference on Decision and Control*, pages 388–391, Tampa Bay, FL, December 1989.

[40] K. Glover and J. Doyle. State-space formulae for all stabilizing controllers that satisfy an $\mathcal{H}_\infty$ norm bound and relations to risk sensitivity. *Systems & Control Letters*, 11(2):167–172, September 1988.

[41] I. Gohberg, M.A. Kaashoek, and H.J. Woerdeman. A maximum entropy principle in the general framework of the band method. *Journal of Functional Analysis*, 95(2):231–254, February 1991.

[42] R.M. Gray. *Entropy and Information Theory.* Springer-Verlag, New York, 1990.

[43] U. Grenander and G. Szegö. *Toeplitz forms and their applications.* University of California Press, Berkeley, CA, 1958.

[44] S.F. Gull. Some misconceptions about entropy. In B. Buck and V.A. Macaulay, editors, *Maximum entropy in action.* Oxford Science Publications, Oxford, UK, 1991.

[45] A. Halanay and V. Ionescu. *Time-Varying Discrete Linear Systems,* volume 68 of *Operator Theory, Advances and Applications.* Birkhäuser, Basel, 1994.

[46] S. Hara and P.T. Kabamba. On computing the induced norm of sampled-data systems. *IEEE Transactions on Automatic Control,* 38(9):1337–1357, September 1993.

[47] R.V.L. Hartley. Transmission of information. *Bell System Technical Journal,* 7(3):535–563, July 1928.

[48] K.L. Hitz and B.D.O. Anderson. Discrete positive real functions and their applications in system stability. *IEE Proceedings,* 116(1):153–155, January 1969.

[49] K. Hoffman. *Banach Spaces of Analytic Functions.* Prentice-Hall, Englewood Cliffs, NJ, 1962.

[50] P.A. Iglesias. Input-output stability of sampled-data linear time-varying systems. *IEEE Transactions on Automatic Control,* 40(9):1646–1650, September 1995.

[51] P.A. Iglesias. An entropy formula for time-varying discrete-time control system. *SIAM Journal on Control and Optimization,* 34(5):1691–1706, September 1996.

[52] P.A. Iglesias and K. Glover. A state space approach to discrete-time $\mathcal{H}_\infty$ control. *International Journal of Control,* 54(5):1031–1074, November 1991.

[53] P.A. Iglesias and D. Mustafa. A separation principle for discrete time controllers satisfying a minimum entropy criterion. *IEEE Transactions on Automatic Control,* 38(10):1525–1530, October 1993.

[54] P.A. Iglesias, D. Mustafa, and K. Glover. Discrete-time $\mathcal{H}_\infty$ controllers satisfying a minimum entropy criterion. *Systems & Control Letters,* 14(4):275–286, April 1990.

[55] P.A. Iglesias and M.A. Peters. On the induced norms of discrete-time and hybrid time-varying systems. *International Journal of Robust and Nonlinear Control*, 1997. To appear.

[56] V. Ionescu and M. Weiss. The ℓ_2-control problem for time-varying discrete systems. *Systems & Control Letters*, 18(5):371–381, May 1992.

[57] A. Isidori. *Nonlinear Control Systems.* Springer-Verlag, Berlin, 2nd edition, 1989.

[58] E.T. Jaynes. Information theory and statistical mechanics. *Physical Review*, 106(3):620–630, May 1957.

[59] E.W. Kamen, P.P. Khargonekar, and K.R. Poolla. A transfer-function approach to linear time-varying discrete-time systems. *SIAM Journal on Control and Optimization*, 23(4):550–565, July 1985.

[60] E.W. Kamen, P.P. Khargonekar, and A. Tannenbaum. Control of slowly-varying linear systems. *IEEE Transactions on Automatic Control*, 34(12):1283–1285, December 1989.

[61] H.K. Khalil. *Nonlinear Systems.* Prentice-Hall, Upper Saddle River, NJ, 2nd edition, 1996.

[62] J.W. Milnor. *Morse theory.* Princeton University Press, Princeton, NJ, 1963.

[63] J.B. Moore and B.D.O. Anderson. Coping with singular transition matrices in estimation and control stability theory. *International Journal of Control*, 31(3):571–586, March 1980.

[64] D. Mustafa and K. Glover. *Minimum Entropy $\mathcal{H}_\infty$ Control*, volume 146 of *Lecture Notes in Control and Information Sciences.* Springer-Verlag, Berlin, 1990.

[65] A. Papoulis. *Probability, Random Variables, and Stochastic Processes.* McGraw-Hill, New York, 2nd edition, 1984.

[66] S. Parrott. On a quotient norm and the Sz-Nagy Foias lifting theorem. *Journal of Functional Analysis*, 30(3):311–328, December 1978.

[67] R. Ravi, K.M. Nagpal, and P.P. Khargonekar. $\mathcal{H}_\infty$ control of linear time-varying systems: A state-space approach. *SIAM Journal on Control and Optimization*, 29(6):1394–1413, November 1991.

[68] R.M. Redheffer. On a certain linear fractional transformation. *Journal of Mathematics and Physics*, 39:269–286, 1960.

[69] W. Rudin. *Real and Complex Analysis.* McGraw-Hill, New York, 3rd edition, 1987.

[70] W.J. Rugh. *Linear System Theory.* Prentice-Hall, Upper Saddle River, NJ, 2nd edition, 1996.

[71] T.L. Saaty. *Modern nonlinear equations.* Dover, New York, 2nd edition, 1981.

[72] I.W. Sandberg. Linear maps and impulse responses. *IEEE Transactions on Circuits and Systems*, 35(2):201–206, February 1988.

[73] C.E. Shannon. A mathematical theory of communication. *Bell System Technical Journal*, 27(3):379–423, (4):623–656, July, October 1948.

[74] A.L. Shields. Weighted shift operators and analytic function theory. In C. Pearcy, editor, *Topics in Operator Theory*, number 13 in Mathematical Surveys, pages 49–128. American Mathematical Society, Providence, RI, 1974.

[75] A.A. Stoorvogel. The discrete-time $\mathcal{H}_\infty$ control problem with measurement feedback. *SIAM Journal on Control and Optimization*, 29(1):182–202, January 1992.

[76] G. Tadmor. Worst-case design in the time domain: The maximum principle and the standard $\mathcal{H}_\infty$ problem. *Mathematics of Control, Signals and Systems*, 3(4):301–324, 1990.

[77] A.J. van der Schaft. $\mathcal{L}_2$-gain analysis of nonlinear systems and nonlinear state feedback $\mathcal{H}_\infty$ control. *IEEE Transactions on Automatic Control*, 37(6):770–784, June 1992.

[78] A.-J. van der Veen. *Time-varying system theory and computational modeling.* PhD thesis, Delft Technical University, The Netherlands, 1993.

[79] L.Y. Wang and G. Zames. Local-global double algebras for slow $\mathcal{H}_\infty$ adaptation: Part II — Optimization of stable plants. *IEEE Transactions on Automatic Control*, 36(2):143–151, February 1991.

[80] L.Y. Wang and G. Zames. Local-global double algebras for slow $\mathcal{H}_\infty$ adaptation: The case of ℓ^2 disturbances. *IMA Journal of Mathematical Control and Information*, 8(3):287–319, 1991.

[81] J.C. Willems. Least squares statinoary optimal control and the algebraic Riccati equation. *IEEE Transactions on Automatic Control*, 16(6):621–634, December 1971.

[82] D.C. Youla. On the factorization of rational matrices. *IRE Transactions on Information Theory*, 7(3):172–189, July 1961.

[83] L.A. Zadeh. Frequency analysis of variable networks. *Proceedings of the IRE*, 38:291–299, March 1950.

[84] G. Zames. On the input-output stability of time-varying nonlinear feedback systems, part I. *IEEE Transactions on Automatic Control*, 11(2):228–238, April 1966.

[85] G. Zames. Feedback and optimal sensitivity: Model reference transformations, multiplicative seminorms, and approximate inverses. *IEEE Transactions on Automatic Control*, 26(2):301–320, April 1981.

[86] G. Zames and L.Y. Wang. Local-global double algebras for slow $\mathcal{H}_\infty$ adaptation: Part I — Inversion and stability. *IEEE Transactions on Automatic Control*, 36(2):130–142, February 1991.

Notation

The main concepts we use for infinite-dimensional operators are introduced in Chapter 2. In general, we use bold letters to represent infinite-dimensional matrices and vectors. For block-diagonal operators $\boldsymbol{A}$, $\boldsymbol{B}$, $\boldsymbol{C}$ and $\boldsymbol{D}$, the notation

$$\left[\begin{array}{c|c} \boldsymbol{A} & \boldsymbol{B} \\ \hline \boldsymbol{C} & \boldsymbol{D} \end{array}\right] := \left\{ \begin{array}{rcl} \boldsymbol{Z}\boldsymbol{x} & = & \boldsymbol{A}\boldsymbol{x} + \boldsymbol{B}\boldsymbol{w} \\ \boldsymbol{z} & = & \boldsymbol{C}\boldsymbol{x} + \boldsymbol{D}\boldsymbol{w} \end{array} \right.$$

is used for a state-space realization of a system.

Next we will list most of the symbols and notation used throughout the book, as well as all the acronyms.

List of symbols

Symbol	Meaning	page
$\text{diag}\{W_k\}$	diagonal operator with W_k as k^{th} element	16
$\langle \boldsymbol{x}, \boldsymbol{y} \rangle$	inner product of $\boldsymbol{x}$ and $\boldsymbol{y}$	16
$\lVert G \rVert_\infty$	$\mathcal{H}_\infty$ norm of $G(z)$	3
$\lVert G \rVert_2$	$\mathcal{H}_2$ norm of $G(z)$	2
$\lVert \boldsymbol{x} \rVert_2$	ℓ_2 norm of $\boldsymbol{x}$	16
$\lVert \boldsymbol{G} \rVert$	induced ℓ_2 operator norm of $\boldsymbol{G}$	17
$\boldsymbol{G}_{[k]}$	k^{th} shifted subdiagonal of $\boldsymbol{G}$	61
$\widehat{\boldsymbol{G}}(\boldsymbol{W})$	$\mathcal{W}$ transform of $\boldsymbol{G}$ evaluated at $\boldsymbol{W}$	62
M^T	transpose of M	
M^H	Hermitian transpose of M	
$G^\sim$	para-Hermitian conjugate of G	47

$\boldsymbol{W}^*$	adjoint of $\boldsymbol{W}$	16
$\boldsymbol{W} > 0$	$\boldsymbol{W}$ is positive definite	18
$\boldsymbol{W} \geq 0$	$\boldsymbol{W}$ is positive semi-definite	18
$\boldsymbol{W}^{-1}$	inverse of $\boldsymbol{W}$	18
$\boldsymbol{W}^{1/2}$	$\boldsymbol{V} \geq \boldsymbol{0}$ such that $\boldsymbol{W} = \boldsymbol{V}^*\boldsymbol{V}$	20
$\mathcal{B}$	class of bounded operators	16
$\mathcal{B}^{-1}$	$\boldsymbol{W} \in \mathcal{B}^{-1}$ if and only if $\boldsymbol{W}^{-1} \in \mathcal{B}$	18
$\mathcal{C}$	class of bounded causal operators	16
$\mathcal{C}^{-1}$	$\boldsymbol{W} \in \mathcal{C}^{-1}$ if and only if $\boldsymbol{W}^{-1} \in \mathcal{C}$	18
$\mathbb{C}$	the set of complex numbers	
$\mathcal{E}$	mathematical expectation	
$\boldsymbol{E}(\boldsymbol{G}, \gamma)$	entropy of the causal operator $\boldsymbol{G}$	48
$\boldsymbol{E}_a(\boldsymbol{H}, \gamma)$	entropy of the anti-causal operator $\boldsymbol{H}$	57
$\mathcal{F}_\ell(\boldsymbol{G}, \boldsymbol{K})$	linear fractional transformation of $\boldsymbol{G}$ and $\boldsymbol{K}$	20
$\boldsymbol{H}_h$	hold operator with holding time h	30
$I_c(G, \gamma)$	entropy of the continuous-time LTI system G	130
$I_d(G, \gamma)$	entropy of the discrete-time LTI system G	47
ℓ_2	set of square summable signals	16
$\mathcal{L}_2$	set of square integrable signals	129
$\mathcal{M}$	class of bounded memoryless operators	16
$\mathcal{M}^{-1}$	$\boldsymbol{W} \in \mathcal{M}^{-1}$ if and only if $\boldsymbol{W}^{-1} \in \mathcal{M}$	18
$\mathbb{N}$	the set $\{1, 2, \cdots\}$	
$\boldsymbol{P}_k$	projection operator	19
$\boldsymbol{P}_k^\perp$	orthogonal complement of $\boldsymbol{P}_k$	19
$\boldsymbol{Q}(\boldsymbol{G})$	quadratic cost of the causal operator $\boldsymbol{G}$	74
$\boldsymbol{Q}_a(\boldsymbol{H})$	quadratic cost of the anti-causal operator $\boldsymbol{H}$	79
$\mathbb{R}$	the set of real numbers	
$\boldsymbol{R}(\boldsymbol{G}, \theta)$	LEQG cost associated with the causal operator $\boldsymbol{G}$	86
$\boldsymbol{S}_h$	sampling operator with sampling time h	30
$\mathbb{Z}$	the set of integers	
$\boldsymbol{Z}$	forward shift operator	19
δ_k	Kronecker delta	2
Δ_k	$\boldsymbol{P}_k - \boldsymbol{P}_{k-1}$	154
Ω	time-reverse operator	24
$\rho(\boldsymbol{T})$	spectral radius of the operator $\boldsymbol{T}$	21

List of acronyms

Acronym	Meaning
DF	Disturbance Feedforward
FC	Full Control
FI	Full Information
LEQG	Linear Exponential Quadratic Gaussian
LQG	Linear Quadratic Gaussian
LTI	Linear Time-Invariant
LTV	Linear Time-Varying
OE	Output Estimation
UES	Uniformly Exponentially Stable

Index

adjoint, 16
almost eigenvalue, 124–126
almost eigenvector, 124, 125
auxiliary cost, 139

bounded real lemma, 30

Cholesky factorization, 20
co-inner operator, 18, 109
co-spectral factorization, 20, 28, 57, 59, 60
conditional entropy, 55
contractive operator, 99, 151
coprime factorization, 98
cost function
    $\mathcal{H}_\infty$ control, 4, 70, 72, 143–145
    $\mathcal{H}_2$ control, 2
    LEQG, 8, 85, 87, 150
    quadratic, 73–76, 79, 87, 92, 146–149
covariance extension problem, 10, 11

detectable, *see* uniform detectability
diffeomorphism, 67
disturbance feedforward problem, 110

eigenvalue, 22, 50, 72, 169
entropy
    anti-causal, 57, 59, 73, 83
    at time k, 48, 50
    average, 57, 83, 87, 109, 121
    continuous-time, 66, 130
    discrete-time, 47, 53, 54
    finite horizon, 93, 150
    sampled system, 140
    time-invariant, 7, 47, 53, 54, 93, 130, 136
entropy rate, 10, 57

finite horizon, 89, 91, 149
forward shift operator, 19
four block problem, 6, 11
full control problem, 106
    all closed-loop systems, 108
full information problem, 101
    all closed-loop systems, 103

Hamilton-Jacobi equation, 65, 66
$\mathcal{H}_\infty$ control, 2, 11
hold operator, 30, 37, 139
$\mathcal{H}_2$ control, 2, 7
hybrid systems, 30, 140

information theory, 8, 9, 54
inner operator, 18, 97, 102
inner product, 16
integral operator, 129, 167

invertible operator, 18

joint entropy, 57

Karhunen-Loève theorem, 168
kernel, 129, 167

lifting operator, 42
linear exponential quadratic Gaussian (LEQG), 8, 85, 149
linear fractional transformation, 5, 20, 151
linear quadratic Gaussian (LQG), *see* $\mathcal{H}_2$ control
Lyapunov equation, 147

matrix extension problem, 10, 11
maximum entropy principle, 8–11
Mercer's theorem, 167
minimum entropy control, 99
  disturbance feedforward, 114
  full control, 109
  full information, 105
  output estimation, 117
  output feedback, 121
  separation principle, 122
Morse's lemma, 65

non-linear systems, 64
norm
  operator, 17, 26–28, 129
  signal, 16, 129

operator, 16
  anti-causal, 16
  bounded, 16
  causal, 16
  diagonal, 16
  memoryless, 16, 48, 74
output estimation problem, 116
output feedback problem, 118

para-Hermitian conjugate, 47, 136
Poisson's integral formula, 47, 136
Popov-Belevitch-Hautus, 123
positive definite, 18
positive semi-definite, 18
probability density function, 8–10, 55, 168
projection operator, 19, 70, 143

quadratic cost, *see* cost function

random variable, 8
  continuous-type, 55
  discrete-type, 54
  Gaussian, 9, 168
realization, 21, 27, 28, 53, 58, 64, 75, 129, 137, 147
Redheffer's lemma, 97
Riccati equation, 53, 58
  algebraic, 6, 45, 54, 66
  bounded real lemma, 30
  co-spectral factorization, 28
  computation, 44
  differential, 137
  $\mathcal{H}_\infty$ control, 101, 106, 120
  hybrid systems, 36, 41, 44
  spectral factorization, 27
risk sensitive control, *see* linear exponential quadratic Gaussian

sampled-data systems, *see* hybrid systems
sampling operator, 30, 32, 139
Shannon entropy, 8
slowly time-varying systems, 37, 45, 63
small gain theorem, 3

spectral factorization, 19, 27, 48, 58, 131, 137
spectral radius, 21
stabilizable, *see* uniform stabilizability
state-space, *see* realization
Stein equation, 23, 73

thermodynamics, 8
time-reverse operator, 24, 59
Toeplitz matrix, 10, 11
Toeplitz operator, 34, 53, 60, 62

uniform
- asymptotic stability, 21
- detectability, 23
- exponential stability, 21, 24, 129
- observability, 124
- stabilizability, 23

unitary, 22, 131

white noise, 80, 87, 93, 146, 150
- at time k, 74, 79, 85

Wiener-Hopf-Kalman, 2

Youla parameterization, 5

Systems & Control: Foundations & Applications

Founding Editor
Christopher I. Byrnes
School of Engineering and Applied Science
Washington University
Campus P.O. 1040
One Brookings Drive
St. Louis, MO 63130-4899
U.S.A.

Systems & Control: Foundations & Applications publishes research monographs and advanced graduate texts dealing with areas of current research in all areas of systems and control theory and its applications to a wide variety of scientific disciplines.

We encourage the preparation of manuscripts in TEX, preferably in Plain or AMS TEX—LaTeX is also acceptable—for delivery as camera-ready hard copy which leads to rapid publication, or on a diskette that can interface with laser printers or typesetters.

Proposals should be sent directly to the editor or to: Birkhäuser Boston, 675 Massachusetts Avenue, Cambridge, MA 02139, U.S.A.

Estimation Techniques for Distributed Parameter Systems
H.T. Banks and K. Kunisch

Set-Valued Analysis
Jean-Pierre Aubin and Hélène Frankowska

Weak Convergence Methods and Singularly Perturbed Stochastic Control and Filtering Problems
Harold J. Kushner

Methods of Algebraic Geometry in Control Theory: Part I
Scalar Linear Systems and Affine Algebraic Geometry
Peter Falb

H^∞-Optimal Control and Related Minimax Design Problems
Tamer Başar and Pierre Bernhard

Identification and Stochastic Adaptive Control
Han-Fu Chen and Lei Guo

Viability Theory
Jean-Pierre Aubin

Representation and Control of Infinite Dimensional Systems, Vol. I
A. Bensoussan, G. Da Prato, M. C. Delfour and S. K. Mitter

Representation and Control of Infinite Dimensional Systems, Vol. II
A. Bensoussan, G. Da Prato, M. C. Delfour and S. K. Mitter

Mathematical Control Theory: An Introduction
Jerzy Zabczyk

H_∞-Control for Distributed Parameter Systems: A State-Space Approach
Bert van Keulen

Disease Dynamics
Alexander Asachenkov, Guri Marchuk, Ronald Mohler, Serge Zuev

Theory of Chattering Control with Applications to Astronautics, Robotics, Economics, and Engineering
Michail I. Zelikin and Vladimir F. Borisov

Modeling, Analysis and Control of Dynamic Elastic Multi-Link Structures
J. E. Lagnese, Günter Leugering, E. J. P. G. Schmidt

First Order Representations of Linear Systems
Margreet Kuijper

Hierarchical Decision Making in Stochastic Manufacturing Systems
Suresh P. Sethi and Qing Zhang

Optimal Control Theory for Infinite Dimensional Systems
Xunjing Li and Jiongmin Yong

Generalized Solutions of First-Order PDEs: The Dynamical Optimization Process
Andreĭ I. Subbotin

Finite Horizon H_∞ and Related Control Problems
M. B. Subrahmanyam

Control Under Lack of Information
A. N. Krasovskii and N. N. Krasovskii

H^∞-Optimal Control and Related Minimax Design Problems
A Dynamic Game Approach
Tamer Başar and Pierre Bernhard

Control of Uncertain Sampled-Data Systems
Geir E. Dullerud

Robust Nonlinear Control Design: State-Space and
Lyapunov Techniques
Randy A. Freeman and Petar V. Kokotović

Adaptive Systems: An Introduction
Iven Mareels and Jan Willem Polderman

Sampling in Digital Signal Processing and Control
Arie Feuer and Graham C. Goodwin

Ellipsoidal Calculus for Estimation and Control
Alexander Kurzhanski and István Vályi

Minimum Entropy Control for Time-Varying Systems
Marc A. Peters and Pablo A. Iglesias